Essentials of a Successful Biostatistical Collaboration

Chapman & Hall/CRC Biostatistics Series

Editor-in-Chief

Shein-Chung Chow, Ph.D., Professor, Department of Biostatistics and Bioinformatics, Duke University School of Medicine, Durham, North Carolina

Series Editors

Byron Jones, Biometrical Fellow, Statistical Methodology, Integrated Information Sciences, Novartis Pharma AG, Basel, Switzerland

Jen-pei Liu, Professor, Division of Biometry, Department of Agronomy, National Taiwan University, Taipei, Taiwan

Karl E. Peace, Georgia Cancer Coalition, Distinguished Cancer Scholar, Senior Research Scientist and Professor of Biostatistics, Jiann-Ping Hsu College of Public Health, Georgia Southern University, Statesboro, Georgia

Bruce W. Turnbull, Professor, School of Operations Research and Industrial Engineering, Cornell University, Ithaca, New York

Published Titles

Published Titles

Benefit-Risk Assessment Methods in Medical Product Development: Bridging Qualitative and Quantitative Assessments
Qi Jiang and Weili He

Biosimilars: Design and Analysis of Follow-on Biologics
Shein-Chung Chow

Biostatistics: A Computing Approach
Stewart J. Anderson

Cancer Clinical Trials: Current and Controversial Issues in Design and Analysis
Stephen L. George, Xiaofei Wang, and Herbert Pang

Causal Analysis in Biomedicine and Epidemiology: Based on Minimal Sufficient Causation
Mikel Aickin

Clinical and Statistical Considerations in Personalized Medicine
Claudio Carini, Sandeep Menon, and Mark Chang

Clinical Trial Data Analysis using R
Ding-Geng (Din) Chen and Karl E. Peace

Clinical Trial Methodology
Karl E. Peace and Ding-Geng (Din) Chen

Computational Methods in Biomedical Research
Ravindra Khattree and Dayanand N. Naik

Computational Pharmacokinetics
Anders Källén

Confidence Intervals for Proportions and Related Measures of Effect Size
Robert G. Newcombe

Controversial Statistical Issues in Clinical Trials
Shein-Chung Chow

Data Analysis with Competing Risks and Intermediate States
Ronald B. Geskus

Data and Safety Monitoring Committees in Clinical Trials
Jay Herson

Design and Analysis of Animal Studies in Pharmaceutical Development
Shein-Chung Chow and Jen-pei Liu

Design and Analysis of Bioavailability and Bioequivalence Studies, Third Edition
Shein-Chung Chow and Jen-pei Liu

Design and Analysis of Bridging Studies
Jen-pei Liu, Shein-Chung Chow, and Chin-Fu Hsiao

Design & Analysis of Clinical Trials for Economic Evaluation & Reimbursement: An Applied Approach Using SAS & STATA
Iftekhar Khan

Design and Analysis of Clinical Trials for Predictive Medicine
Shigeyuki Matsui, Marc Buyse, and Richard Simon

Design and Analysis of Clinical Trials with Time-to-Event Endpoints
Karl E. Peace

Design and Analysis of Non-Inferiority Trials
Mark D. Rothmann, Brian L. Wiens, and Ivan S. F. Chan

Difference Equations with Public Health Applications
Lemuel A. Moyé and Asha Seth Kapadia

DNA Methylation Microarrays: Experimental Design and Statistical Analysis
Sun-Chong Wang and Arturas Petronis

DNA Microarrays and Related Genomics Techniques: Design, Analysis, and Interpretation of Experiments
David B. Allison, Grier P. Page, T. Mark Beasley, and Jode W. Edwards

Dose Finding by the Continual Reassessment Method
Ying Kuen Cheung

Dynamical Biostatistical Models
Daniel Commenges and Hélène Jacqmin-Gadda

Elementary Bayesian Biostatistics
Lemuel A. Moyé

Empirical Likelihood Method in Survival Analysis
Mai Zhou

Essentials of a Successful Biostatistical Collaboration
Arul Earnest

Published Titles

Published Titles

Sample Size Calculations in Clinical Research, Second Edition
Shein-Chung Chow, Jun Shao, and Hansheng Wang

Statistical Analysis of Human Growth and Development
Yin Bun Cheung

Statistical Design and Analysis of Clinical Trials: Principles and Methods
Weichung Joe Shih and Joseph Aisner

Statistical Design and Analysis of Stability Studies
Shein-Chung Chow

Statistical Evaluation of Diagnostic Performance: Topics in ROC Analysis
Kelly H. Zou, Aiyi Liu, Andriy Bandos, Lucila Ohno-Machado, and Howard Rockette

Statistical Methods for Clinical Trials
Mark X. Norleans

Statistical Methods for Drug Safety
Robert D. Gibbons and Anup K. Amatya

Statistical Methods for Healthcare Performance Monitoring
Alex Bottle and Paul Aylin

Statistical Methods for Immunogenicity Assessment
Harry Yang, Jianchun Zhang, Binbing Yu, and Wei Zhao

Statistical Methods in Drug Combination Studies
Wei Zhao and Harry Yang

Statistical Testing Strategies in the Health Sciences
Albert Vexler, Alan D. Hutson, and Xiwei Chen

Statistics in Drug Research: Methodologies and Recent Developments
Shein-Chung Chow and Jun Shao

Statistics in the Pharmaceutical Industry, Third Edition
Ralph Buncher and Jia-Yeong Tsay

Survival Analysis in Medicine and Genetics
Jialiang Li and Shuangge Ma

Theory of Drug Development
Eric B. Holmgren

Translational Medicine: Strategies and Statistical Methods
Dennis Cosmatos and Shein-Chung Chow

Chapman & Hall/CRC Biostatistics Series

Essentials of a Successful Biostatistical Collaboration

Arul Earnest

Monash University
Australia

CRC Press
Taylor & Francis Group
Boca Raton   London   New York

CRC Press is an imprint of the
Taylor & Francis Group, an **informa** business

A CHAPMAN & HALL BOOK

CRC Press
Taylor & Francis Group
6000 Broken Sound Parkway NW, Suite 300
Boca Raton, FL 33487-2742

First issued in paperback 2020

ISBN 13: 978-0-367-57444-4 (pbk)
ISBN 13: 978-1-4822-2698-0 (hbk)

Version Date: 20160808

Library of Congress Cataloging-in-Publication Data

Names: Earnest, Arul, author.
Title: Essentials of a successful biostatistical collaboration / Arul Earnest.
Description: Boca Raton : Taylor & Francis, 2016. | Series: Chapman &
Hall/CRC biostatistics series | Includes bibliographical references and
index.
Identifiers: LCCN 2016014111 | ISBN 9781482226980 (alk. paper)
Subjects: LCSH: Clinical medicine--Research--Statistical methods. | Sampling
(Statistics)
Classification: LCC R853.S7 E27 2016 | DDC 616.0072/7--dc23
LC record available at https://lccn.loc.gov/2016014111

Visit the Taylor & Francis Web site at
http://www.taylorandfrancis.com

and the CRC Press Web site at
http://www.crcpress.com

I would like to show appreciation to my wife, Josephine, and daughters, Megan and Emma, for their support. I also dedicate the book to my mum, Mariammah, and my late father, Stephen Monickam, for their unfailing belief in the value of education.

Contents

Preface

The biostatistics profession is relatively new in the field of medicine. Most biostatisticians learn their tools of the trade through on-job training, and they either pick up the skills on the job through experience gained from a large number of collaborative projects or learn from an experienced mentor if available. Current resources on 'consulting' for biostatisticians is targeted at statistical consulting, which is different from a 'collaboration', and they include resources that provide examples from a diverse range of applications with few medical examples, as they are not specifically targeted at biostatisticians (Daniel, 1969; Demming, 1965; Hahn and Hoerl, 1998), or contain a collection of papers on statistical consulting in diverse areas such as finance, science, medicine and marketing with each study exemplifying a sample of issues, including crossover designs and ethical issues with clinical trials (Hand and Everitt, 1987) and case studies, including battery failure analysis, job promotion discrimination, lost mail analysis and food experiments (J. and Mc Dougall, 2010). A comprehensive resource covering all the aspects of collaboration with clinicians, especially writing for a publication or a grant proposal, oral and written communication and presentation skills, and managing projects and collaborators is currently lacking. Those that endeavour to cover communication aspects, for instance, unfortunately do not cover the technical and methodological aspects, which are equally important in collaboration (Derr, 2000) or focus on writing skills for reports, not publications (J. and Mc Dougall, 2010), the latter being more critical in a collaboration. This book will provide valuable insights for a practising biostatistician on 'hard skills' like methodological concepts revolving around study designs, form and database design, statistical analysis plan (SAP) and sample size calculations and non-methodological 'soft skills', such as how best to communicate with clinicians, as well as the best practices to adopt in terms of project, time and data management, writing skills and managing relationships with collaborators.

A unique aspect of this book is the dissemination of results from a survey among practising biostatisticians from a diverse background of education and working institutions, to gauge their responses on important issues such as problems ever faced in collaborating with clinicians, training on collaboration/consultation skills, frequency of performing selected tasks as a biostatistician, issues to address and important skills to gain in order to enhance greater collaborations and so on. The results of this survey are presented in Chapter 11, and they also provide context upon which many of the other chapters are developed, including communication skills and project management expertise. In the same chapter, we also present 'views from the other side', that is, qualitative interviews with experienced clinician researchers who have worked with a number of biostatisticians in the past and have published extensively in a range of clinical fields, including infectious disease, rheumatology and psychiatry. Thoughts on how biostatisticians have added value to their research projects, challenges faced in collaborating with biostatisticians and specific knowledge or skill sets that they think biostatisticians should acquire to raise the standard of collaborative input are presented.

Chapter 8 addresses one major current gap in the training needs for academic biostatisticians, namely best practices in project management. Topics such as the creation of a project file to keep track of and manage important collaborative projects, guidelines on the naming and organisation of files and folders, database security and confidentiality, ensuring consistency and reproducibility in the statistical analysis, tips for working on multiple collaborations successfully as well as choosing an ideal mentor and working on a successful relationship with the mentor are discussed.

Some biostatisticians work in isolation in their institutions with few or even no colleagues to check with at their workplace, and this book intends to serve as a guide that they can use in their everyday work, and hopefully shorten their learning curve. The book will also be an ideal resource for postgraduate biostatistics and epidemiology students, to prepare them for a career as a collaborating biostatistician in an academic institution/hospital. The ultimate objective is to raise the quality and quantity of collaborative research that biostatisticians are involved in. It has been generally recognised that globally there is a shortage of academic biostatisticians, especially in hospitals, medical schools and research institutions, and this has been attributed to the increased recognition of the contribution statistics has made to all fields of health research, insufficient training opportunities in applied and medical statistics and inadequate supply of capable interest to make a career in this field (Pocock, 1995). In addition to the existing biostatisticians based around the various hospitals and research institutions around the world, the number of new biostatisticians entering the field is thus likely to increase, and this book will serve as a useful resource for them.

In terms of academic level, it is anticipated that this book will appeal to readers with at least a bachelor's degree in a quantitative field, including statistics. Those with a postgraduate degree in biostatistics and epidemiology will also benefit. This book can be used as a teaching module for postgraduate courses such as MSc and PhD in biostatistics, epidemiology and public health. This book is primarily targeted for teaching purposes, and intended for biostatisticians, epidemiologists and any quantitative scientists working in academic institutions/hospitals and who are involved in collaborative research with clinicians or other researchers.

This book should be useful for clinicians who are involved in some form of clinical and health services research and public health work, and who would like to collaborate with a biostatistician. It should serve its purpose of informing how an effective working relationship can be fostered between the two groups. It is also possible that the book can be used as a didactic teaching tool for clinical research programmes designed for clinicians who are interested in a career in research.

This book is organised along the following framework: the first five chapters provide a summary of the essential methodological and analytical skills that a practising biostatistician needs to have in order to perform his/her roles and functions as a statistical collaborator effectively, particularly within an academic or hospital setting. This is naturally organised with the design of the study presented first (e.g. observational and randomised controlled trial, data collection form and database design, sample size and power calculations) followed by the chapter on SAP. The SAP chapter is exemplified with clinical examples using the Stata statistical software, with an aim towards introducing to readers some of the common statistical techniques encountered in the clinical research setting. This will provide guidance to the biostatistician on the appropriate statistical tests to employ depending on the aims and hypotheses. Although Stata has been used to exemplify the statistical technique and interpretation of the output with the clinical examples provided, the key learning points are applicable to other software as well. This chapter contrasts with current resources that aim to present the use of Stata to analyse selected projects and presents a brief description of the data, model and exercises provided at the end of each chapter (Rabe-Hesketh and Everitt, 2000) and another which is targeted at data analysts with guidelines including those on organising data and performing and reporting analysis using programmes written in Stata (Long, 2009).

Chapters 6 through 9 encompass the non-methodological 'soft skills' required in collaborations, namely effective communication, effective writing, how to successfully manage collaboration, good practices in project management and so on. Chapter 10 highlights the 'how not to', that is, the pitfalls in the design, analysis and presentation of data. This chapter endeavours to provide biostatisticians with knowledge on the critical snares in the various stages of the research collaboration cycle. Once again, the focus here is on health care-related examples and projects relevant for the practising biostatistician, compared to other resources aimed towards the discussion of more general statistical errors in hypothesis testing, modelling and interpretation of reports (Good and Hardin, 2001). The final chapter provides results from a survey among practising biostatisticians as well as interviews with experienced clinicians from a range of medical specialties, gauging their views on the collaboration process. The main objective is to present views from biostatisticians with a wide range of experience and expertise on the collaboration process. Another common thread underlying in all the chapters is the provision of 'questions to ask clinicians', which are useful checklists for biostatisticians to bring to the table when meeting their collaborators. These documents should prove useful when discussing with the collaborator about the study design, sample size calculation or when deliberating on the SAP. In addition to the many clinical examples used in the book to exemplify statistical and methodological ideas and concepts, the chapters also end off with a real-life collaborative project, where possible. Valuable lessons are then drawn from the use of actual collaborative projects.

The key features of this book are as follows:

1. Provides an overview of common functions expected from a collaborating biostatistician
2. Identifies tools and resources that a biostatistician needs to work effectively
3. Uses selected published articles from the author's work with clinicians' to highlight key statistical techniques and learning points
4. Asserts best practices in terms of project management and use of gold standard templates in the design and reporting of studies
5. Presents technical statistical concepts in simpler terms that can be conveyed to clinicians
6. Employs actual interviews with biostatisticians and clinicians from a broad spectrum of medical specialities to convey key learning points regarding collaboration with biostatisticians

In this book, statisticians and biostatisticians have been used interchangeably, and they represent the same potential audience who may benefit from this book. In addition, the book does not intend to cover the wide range of statistical applications in the health care field, as there are already books available that cover specific topics in more detail (Altman, 1991; Armitage et al., 2002; Pocock, 2000). The statistical techniques described in the book are based on the more common clinical examples a biostatistician usually encounters in the hospital setting. Therefore, it is not possible to cover all the statistical techniques, e.g. population health methods, such as statistical process control charts, and survey data analysis, the design and analysis issues related to *in vitro* and *in vivo* experiments, the design and data analysis issues related to omics data and issues related to developing predictive models in this book.

This book was motivated by the author's personal journey as a practising collaborative biostatistician for more than 15 years. Apart from obtaining a degree and two postgraduate degrees from three different countries including the United Kingdom, Australia and Singapore, the author also has experience working in various academic and hospital settings in Australia and Singapore, from which real-life examples described in this book are drawn from. This includes selected examples from more than 125 publications in peer-reviewed medical journals, 2 book chapters, 13 successful grant applications, more than 130 conference proceedings, as well as more than 50 invited talks and presentations that he has been involved in over the past 15 years. The author has also taught in formal courses on epidemiology, health statistics, multi-variate analysis, grantmanship and statistical software classes. In addition to mentoring statisticians, clinicians in residency programmes and clinician-researchers, he has also taught successive batches of medical students and master's in public health students both in Australia and in Singapore. The author has reviewed numerous grant proposals and protocols while serving as a regular member of local review panel for grants submitted to the National Medical Research Council in Singapore, as well as the Institutional Review Board of the National Healthcare Group.

The author was previously an academic editor of the editorial board of *PLoS ONE*, a non-profit, open-access, online publication that reports on original research from all disciplines within science and medicine, and regularly reviews articles for a number of journals including *Journal of American Medical Association* (JAMA); *BMC Health Services Research, Journal of Urban Health, Annals of the Academy of Medicine, Singapore, Australian Journal of Rural Health, International Journal of Infectious Diseases, Latin American Applied Research – An International Journal*, and *Medical Journal of Australia*. The reviews have provided useful information that has been presented in this book. He has also been invited to sit in the judging panel in numerous competitions, including the Singapore General Hospital Annual Scientific Meeting, Singapore Health and Biomedical Congress and Pitch for Fund, Medicine Academic Clinical Program at Singapore General Hospital, where he has had the opportunity to critically evaluate various clinical research projects.

From April 2011 to August 2014, the author was the director of the Centre for Quantitative Medicine at the Duke-NUS Graduate Medical School in Singapore. During this time, he was responsible for the organisation and allocation of faculty time to various activities, such as teaching of medical students, clinician research programmes such as the Master of Clinical Investigation, the Khoo Scholars and the Research Development Seminars. During the same period, he also worked with senior clinicians, hospital administrators and chief executive officers from various hospitals and research institutions, to further extend the reach of biostatistical collaborations of the Centre to other health care institutions such as the National Neuroscience Institute and the SingHealth Polyclinics, Singapore. He was also involved in the organisation and successful execution of two workshops designed to help equip biostatisticians with the skills necessary to collaborate successfully with clinicians. Input for this book was also obtained from fellow biostatisticians and clinician collaborators through a structured self-administered survey and qualitative interviews. A number of collaborators (both biostatisticians and clinicians) from the United States, Australia and Singapore have also looked at the draft chapters and provided critical input, and they have been recognised in the 'Acknowledgements' section.

References

Altman, D. G. (1991). *Practical Statistics for Medical Research*. London: Chapman & Hall/CRC Press.

Armitage, P., Berry, G., & Matthews, J. (2002). *Statistical Methods in Medical Research* (4th ed.). Oxford: Blackwell Science.

Cabrera, J., & Mc Dougall, A. (2010). *Statistical Consulting*. New York: Springer-Verlag.

Daniel, C. (1969). Some general remarks on consulting in statistics. *Technometrics, 11*, 241–245.

Demming, W. E. (1965). Principles of professional statistical practice. *Ann Math Stat, 36*, 1883–1900.

Derr, J. (2000). *A Guide to Effective Communication* (1st ed.). Pacific Grove, CA: Duxbury Press.

Hahn, G., & Hoerl, R. (1998). Key challenges for statisticians in business and industry. *Technometrics, 40*, 195–200.

Hand, D. J., & Everitt, B. S. (1987). *The Statistical Consultant in Action*. Cambridge: Cambridge University Press.

Long, J. S. (2009). *The Workflow of Data Analysis Using Stata*. College Station, TX: Stata Press.

Pocock, S. J. (1995). Life as an academic medical statistician and how to survive it. *Statistics in Medicine, 14*, 209–222.

Pocock, S. J. (2000). *Clinical Trials – A Practical Approach*. Chichester: John Wiley & Sons.

Rabe-Hesketh, S., & Everitt, B. S. (2000). *A Handbook of Statistical Analyses using Stata* (2nd ed.). Boca Raton, FL: Chapman & Hall/CRC Press.

Acknowledgements

I thank the following people for their comments and suggestions that have significantly improved the quality of this book: Professor Augustus John Rush (emeritus professor, the National University of Singapore, Singapore); Professor John Carlin (the University of Melbourne, Melbourne, Victoria, Australia); Professor Cheung Yin Bun (the Duke-NUS Graduate Medical School, Singapore); Professor Nick Paton (the National University Health System, Singapore); Professors Andrew Forbes and Rory Wolfe (Monash University, Melbourne, Australia); Associate Professor Tan Say Beng (the National Medical Research Council, Singapore); Professor Annelies Wilder-Smith and Associate Professor Denny Meyer (Swinburne University of Technology, Melbourne, Victoria, Australia); Associate Professor Stephane Heritier (Monash University, Melbourne, Australia); Associate Professor G. Shantakumar, Assistant Professor Bibhas Chakraborty, Assistant Professor Nivedita Nadkarni, Assistant Professor Benjamin Haaland, Assistant Professor John Carson Allen, Sanchalika Acharyya, Nurun Nisa De Souza, Mihir Gandhi, Gita Krishnaswamy and Taara Madhavan (all from the Duke-NUS Graduate Medical School, Singapore); and Dr. Nagaendran Kandiah (the National Neuroscience Institute, Singapore). I thank Sivagami from the Centre for Quantitative Medicine, the Duke-NUS Graduate Medical School, Singapore, for helping to key in the data of the survey, for which results are reported in Chapter 11. In addition, I am grateful to Professor Nick Paton (the National University Health System, Singapore), Professor John McNeil (Monash University, Melbourne, Australia) and Professor Augustus John Rush (emeritus professor, the National University of Singapore, Singapore), Associate Professor Geoff Morgan (Sydney Medical School, the University of Sydney, Sydney, New South Wales, Australia) and Dr. Leong Khai Pang (Tan Tock Seng Hospital, Singapore) for providing interviews and sharing their invaluable experience on working with biostatisticians with diverse expertise in a wide range of projects.

Author

Arul Earnest is currently an associate professor of biostatistics at Monash University Department of Epidemiology and Preventive Medicine and an adjunct professor at Swinburne University of Technology, both in Melbourne, Victoria, Australia. Previously, he was the director of the Centre for Quantitative Medicine at the Duke-NUS Graduate Medical School in Singapore, a collaboration between Duke University in North Carolina, United States, and the National University of Singapore. He graduated from London School of Hygiene and Tropical Medicine, London, in 2002, with an MSc in medical statistics and a PhD in statistics from Queensland University of Technology in 2010. His current research interest is in Bayesian spatio-temporal models. For more than 15 years, Earnest has provided consultative and collaborative methodological input to various collaborators and secured several grants, including a large Australian Research Council grant, as well as a National Medical Research Council grant from Singapore. He was awarded the status of Chartered Statistician by the Royal Statistical Society in London in 2003 and has received several research awards, including the recent University of Sydney International Research Collaboration Award in 2013 and the SingHealth Partners in Education Award: RISE award for mentors and teachers of resident clinicians. He has been in several organising committees for national as well as international conferences, including the International Society of Bayesian Analysis conference in 2008.

1

Observational Study Designs

1.1 Introduction

This chapter reviews issues that may confront a biostatistician in relation to observational studies, which are the most common study designs used in the clinical research setting. A unique feature of this chapter is the employment of a single clinical question to describe the important features and issues revolving around the different observational study designs. This will hopefully lead the readers to better understand the comparative features between them. It is not the aim of this chapter to provide a comprehensive summary on the topic of study design, but rather to encapsulate the key elements that need to be discussed during initial meetings with the collaborator when deliberating on the study design (e.g. through a series of questions to ask of collaborators). Using exemplar clinical questions and tools such as 'developing the aims and hypothesis,' the biostatistician can provide critical input on the choice of the study design by helping the collaborator to carefully define the research question. As the biostatistician may sometimes be overwhelmed during initial conversations with the clinician, a list of 'Questions to ask clinicians' when deliberating on the choice of the study design is also provided and discussed. Common mistakes made in the design of observational studies are examined in this chapter, so that these can be avoided when working with the collaborator on the study design features, as it is difficult to make changes to the study design after the study has been completed. Links to online resources and videos are also provided for readers to obtain further information on study designs.

1.2 Comparative Features between Cohort, Case–Control and Cross-Sectional Studies

Cohort, case–control and cross-sectional studies are collectively known as observational studies, where the investigators or clinicians do not necessarily manipulate the way exposure and outcome variables occur. One may employ different study designs to answer the same research question. Each study design has different features, advantages as well as disadvantages, and this can be exemplified with the following research question: Is solar exposure associated with skin cancer among Australians?

1.2.1 Cross-Sectional Study Design

One may select a simple random sample of Australians, drawn from the various housing suburbs. A surrogate measure of solar exposure (e.g. skin elasticity on the back of the hand) can then be measured and subsequently correlated with the presence or absence of physician-diagnosed skin cancer. In such a cross-sectional survey, typically the information is collected from each subject at a single point in time. One main advantage of this design is that the study can be conducted quickly, as we do not need to wait for the outcome to develop. This study design is also relatively inexpensive as subjects do not need to be followed up by research personnel, and there is no problem with dropouts or loss to follow-ups. In a cross-sectional study, the prevalence of disease or risk factors can also be ascertained fairly quickly. This may be a favourable design for a collaborator who does not have much resource (e.g. clinician with no research funding).

A major disadvantage of a cross-sectional study is that we will be unable to establish causality, specifically through demonstrating a temporal relationship between the exposure and the outcome. For instance, it is possible that subjects who had positive solar exposure had skin cancer diagnosed several years before. Possible sources of bias include sample not being representative of the population of study (due to improper random sampling or exclusion of some groups of people) as well as low response rates (Machin et al., 2007). Random sampling is an important feature of this study design, and Armitage and colleagues provide a useful summary of the various forms of sampling (e.g. systematic, stratified, multi-stage, cluster and mark-recapture sampling techniques) (Armitage et al., 2002). When communicating with the clinician, it is important to highlight the importance of employing random sampling, and to disabuse one common misconception that increasing sample size will solve the problem of representativeness.

1.2.2 Case–Control Study Design

An alternative approach to answer the research question posed in Section 1.2 is to employ a case–control study design. In relation to the skin cancer example, one might then recruit patients diagnosed with skin cancer (cases) from the Alfred Hospital in Melbourne, Victoria, Australia, over a period of (say) 1 year, and patients diagnosed with other skin conditions, but not skin cancer (as controls) from the same institution over the same time period.

The cases and controls can be matched by key characteristics such as age and gender. The purpose of matching is to reduce confounding, control for confounders that cannot be measured, as well as increase the precision of comparisons between groups by balancing the number of cases and controls at each level of the confounder (Hulley et al., 2007). However, it has been argued that overmatching may result in an underestimation of the effect size (Machin et al., 2007). When discussing issues of matching with the collaborator, the biostatistician needs to evaluate the difficulty in terms of logistics (i.e. will there be sufficient controls with the characteristics to be matched upon?) and whether matching will be done one-to-one or by groups of patients. The patients' lifetime solar exposure can then be determined through a questionnaire, and this variable can then be subsequently compared between the two groups (cases and controls) using appropriate statistical techniques.

One obvious advantage of this study design is that it is useful for studies where cases (outcomes) are relatively rare, such as rare forms of birth defects (e.g. spina bifida), because we are selecting the samples based on the cases. The study can also be quickly designed and carried out within a reasonably short period of time, because one does not have to follow up the participants for lengthy periods of time. One major problem with the case–control design is the potential of recall bias, in particular relating to the exposure variable. In this example, it is difficult for one to quantify their exposure to the sun over the many years prior to the study. The other problem with this study design lies in possible differential misclassification of the exposure variables (i.e. skin cancer patients could well have a heightened awareness due to their disease and may have been acutely collecting data on solar exposure compared to the controls). Another related design is the case–crossover design, where cases act as their own controls over different periods of time. However, these can only be useful for short-term effects of intermittent exposures and not suitable for outcomes which are not reversible, for example, skin cancer. It is also difficult to establish the incidence or prevalence of the disease with this study design. These issues should be highlighted to the clinician when considering the case–control design for this study.

1.2.3 Cohort Study Design

The cohort study is the strongest study design within observational studies in terms of establishing causality. In the context of our study looking at the association between solar exposure and skin cancer, a sample of individuals without skin cancer is firstly recruited (e.g. those aged 16–45 from the community). Their past and ongoing personal solar exposure is then documented (e.g. through a diary or questionnaire on work and outdoor activities). The cohort is then followed up for a long period of time (e.g. 5–10 years) until they develop skin cancer (i.e. ascertained by a physician), die or dropout of the study. The cohort need to be contacted at regular time points to ascertain the outcome status, or this may be elucidated from national cancer registries (where available). When deliberating with the clinician on the possibility of employing a cohort study, the median age of onset of disease, the annual incidence level of disease and the difficulty in capturing the outcome status should be discussed to determine the feasibility of employing this design.

A key feature and advantage of cohort studies is that the exposures are determined before the outcome, and hence provide for greater evidence towards establishing causality. Exposure (here ultraviolet exposure) can be measured with relatively less bias (both recall and misclassification), for example, via a diary documenting their current/past outdoor activities. The cohort can also be regularly contacted during the follow-up period, and updated information on their solar exposure as well as changes in their occupation/outdoor activities can be collected and used to better correlate with the outcome. In a cohort study, a large number of risk factors can be measured and studied (e.g. smoking, occupation, diet, etc.). Cohorts can be both prospective (current) or historical (defined back in time), although the biostatistician should be aware that there may not be much flexibility in measuring the exposure in a historical cohort (e.g. one may not be able to evaluate a new personal wearable gadget that measures UV rays).

The major limitation of the cohort study is that it can incur a high cost (both monetary and manpower as well as other resources). This is due to the longitudinal nature of the study and the fact that the subjects in the study usually need to be followed up for a long period of time. This can be a major hindrance for a clinician who needs to complete his/her study within 2 years (e.g. part of the fellowship or a requirement in a grant), and hence other study designs should be considered. There is also a potential for a large loss to follow-up (e.g. dropouts and deaths). Although this censoring can be addressed by the time to event (survival) statistical analysis models, the implication is that one needs to start with a large sample size in the first place to observe the required number of events at the end of the study. This will have implications in the sample size calculation (discussed in greater detail in Chapter 4) and should be mentioned to the collaborator at the start of the discussion. In the skin cancer example, a cohort study may not be ideal in populations where the outcome is rare, where the incident rate is low.

Sometimes, it is possible to have a nested case–control design crafted out from a larger initial cohort study. This is a useful design for studying predictor variables that can be expensive to measure and are assessed at the end of the study on subjects who develop the outcome during the study (cases) and controls (Hulley et al., 2007). In our example, we can have a larger cohort of healthy individuals followed up for a long period and those diagnosed with skin cancer selected during the follow-up period. A random sample of non-cases can be selected from the cohort as controls, and their UV status recorded, along with perhaps another measure such as an expensive genetic test that may test whether they are predisposed to skin cancer.

1.2.4 Ecological Study Design

Finally, another approach to studying the above question is to analyse data at an aggregate level, instead of comparing exposure and outcome at the individual level. For instance, one could collect monthly data on UV radiation and skin cancer incidence by countries and employ time series models to establish correlations between the two. Alternatively, one could also collect annual data and compare them across countries, with the unit of analysis being countries. At the outset, this design may appear attractive to the collaborator, because data may readily be collected from external secondary sources and the study can be completed quickly. The biostatistician should caution that there are potential bias arising from ecological fallacy, in which affected individuals in a generally exposed group may not necessarily be the ones exposed to the risk factor (Fletcher & Fletcher, 2005), and thus, relationships at the areal level cannot be generalised or applied to the individual level.

As there can be seasonal trends in the UV data, more complex models may then be needed to analyse data that are autocorrelated over time or geography. Methods such as autoregressive integrated moving average (ARIMA) and conditional autoregressive (CAR) models have typically been used in these instances. For example, in a previously published article (Earnest et al., 2012) using a Poisson model that allowed for autocorrelation and overdispersion in the data, we found weekly mean temperature and mean relative humidity as well as Southern Oscillation Index to be significantly and independently associated with dengue notifications. Seasonal ARIMA (SARIMA), models with periodic components have also been used to predict the temporal trends of monthly TB risk among residents and non-residents in Singapore and detect seasonality (Wah et al., 2014).

Bayesian CAR models have also been used to investigate the geographic variation of bystander cardiopulmonary resuscitation (BCPR) provision and survival to discharge outcomes among residential out-of-hospital cardiac arrest (OHCA) cases, and evaluate this variation with respect to individual and population characteristics, and to identify high-risk residential areas with low relative risk (RR) of BCPR and high RR of OHCA at the development guide plan census tract levels in Singapore (Ong et al., 2014). It should be noted that the time series models and spatial analysis may require fairly sophisticated knowledge of programming skills in Bayesian software as well as information on the statistical models, and the biostatistician should ensure that he/she is able to undertake such analysis on his/her own. Otherwise, he/she should request for more time to pick up the required skills or approach another biostatistician who has experience in the field.

1.3 Selecting the Appropriate Study Design

The choice of the appropriate or optimal study design essentially depends on a few key factors such as the research question and hypothesis, the prevalence of the outcome and exposure variable of interest, and also resources available to the investigator (i.e. time and money). Using our example on skin cancer and solar exposure given in Section 1.2, if the research aim was to establish a causal relationship between skin cancer and exposure to the sun, a case–control study would seem to be the most optimal design as well as cost-effective. However, if the aim was to estimate the prevalence of skin cancer or determine the range of solar exposure among Australians living in a particular suburb in Melbourne, a cross-sectional study may suffice. Alternatively, in order to develop a prognostic index for UV on skin cancer, a cohort study may be required. Sometimes, the collaborator approaches the biostatistician after the study design has been decided and data collected. In those instances, there are very little options for remedial actions, and any possible deficiencies in the study design may be addressed in the analysis to some extent. For instance, in a case–control design that has not been matched by age, multi-variate statistical techniques can be employed to adjust for the effect of age. Alternatively, matching the cases and controls in the analysis stage by using propensity scores can yield unbiased estimates (Pfeiffer & Riedl, 2015).

1.4 Key Questions to Ask a Clinician

1. What is your research question?

   The research question will determine the appropriate study design to apply for the study. For instance, if the research question is 'What is the prevalence of diabetes among patients presenting to a health screening event in a hospital?', then a cross-sectional study design may be appropriate. If the question is phrased as 'Does smoking cause diabetic retinopathy?', then perhaps a case–control or cohort study design would be helpful.

2. How would you describe the specific hypotheses?

   The specific hypothesis will identify the type of study designs that can be used for the study. For example, if the hypothesis involves examining factors associated with changes in quality-of-life scores for patients annually across 5 years, then a longitudinal design such as the cohort study may be appropriate.

3. Are you planning to use existing databases for your research project?

   The use of existing databases may restrict the type of study designs that may be available for the study team. For instance, it may be difficult to conduct a prospective cohort study based on a dataset that has been retrospectively collected. Information is also limited to the variables that are already collected in the dataset, and these variables may subject to bias or measurement error that cannot be easily rectified. This should be highlighted to the collaborator to try and dissuade the use of existing datasets if they are not optimal.

4. Is the study descriptive or inferential in nature?

   Descriptive studies may involve a cross-sectional or ecological study design.

5. Do you need to establish causality in your study?

   In order to establish causality, randomised controlled trials (RCTs) are usually employed. Alternatively, cohort studies may also be employed.

6. Have there been previous attempts to answer this question?

   It may be possible that several smaller studies have been conducted previously to answer the research question, but each study may not be adequately powered to demonstrate any significant effects. Hence, a meta-analysis can be undertaken to synthesise the results of previous studies, rather than to conduct another new study.

7. Are there any logistical constraints (e.g. budget, manpower, sample size and patient accrual)?

   Factors such as small sample size and poor patient recruitment rate may preclude the application of study designs such as a cohort study design or a cluster RCT, as these designs usually require large sample sizes.

8. Is there a deadline by which you would like to complete the study?

   Studies with short or restricted timelines may preclude the use of designs such as prospective cohort or crossover trials. Usually, cross-sectional studies are the most practical (although not the most ideal).

The first and most important information to elicit from the clinician, regardless of whether the purpose is to design the study, calculate the sample size or analyse the data, is 'What is the research question or aim?' The research question must then be translated into a working hypothesis that is very specific, including the WHO (subjects, raters or treaters), HOW (by what means or with what measurements will be taken), WHAT (the 'thing' being studied), WHERE (site/setting), WHEN (in the course of treatment, how long) and BY WHAT METRIC (what is the measure, effect size) (Rush, 2013). We provide some examples on how a statistician can work iteratively with his/her collaborator to help craft out a set of aims and hypotheses that are specific enough to inform on the appropriate design and analysis of the study.

Example 1

Aim. To investigate the effectiveness of aspirin in terms of reducing incidence of recurrent strokes among patients admitted for strokes.

Initial hypothesis. Aspirin will reduce the number of recurrent stroke patients.

As we can observe, the above hypothesis is not specific and lacks important details such as the profile of stroke patients, the follow-up period and the clinically meaningful minimum difference that the investigator wishes to detect.

Revised hypothesis. Among stroke patients aged 40–60 and admitted to the Alfred Hospital from January to July 2010, administering aspirin at 100 mg daily will reduce the incidence of recurrent stroke from 20% to 5% over a follow-up period of 6 months from admission.

Example 2

Aim. To compare the effectiveness of positron emission tomography (PET) scan versus computed tomography (CT) scan in terms of diagnosing neck cancer tumours among elderly patients presenting to a Melbourne hospital.

Initial hypothesis. The sensitivity of PET scans would be better than CT scans when diagnosing patients with neck cancer tumours.

The above hypothesis is not specific as it does not mention how much of an improvement is expected to be seen, as well as the gold standard used to ascertain the presence of neck cancer. It does not mention the profile of the patients, as well as the time when the measurements were obtained.

Revised hypothesis. Among elderly patients (aged 65 and above) who were seen in a tertiary hospital in Melbourne in 2014, the PET scan performed on admission would show an improved sensitivity (99%) compared with a CT scan (95%) when evaluated by blinded assessors at the same time, and compared to the gold standard of a biopsy sample analysed by a pathologist.

With a well-crafted hypothesis, the statistician can then suggest the appropriate study design, as well as the relevant statistical method to analyse the data. For Example 1, a cohort study design might be prescribed as there is follow-up of patients involved in the study. A case–control study would not be appropriate as the purpose is not to compare against a control group, and there may not be sufficient variability in the exposure variable (i.e. prescription of aspirin). For Example 2, if PET and CT scans are routinely performed for patients with neck cancer in that particular hospital, then a case–control study may be appropriate. The cases can be neck cancer cases, and controls selected from the same hospital among those without the disease in the same time period. If the scans are not routinely performed, then the study can be designed prospectively as a cohort study with appropriate protocols set in place to ensure all patients suspected with neck cancer receive PET and CT scans and are assessed for neck cancer using histology samples. The biostatistician should also find out the prevalence of exposure and outcome in his or her population of interest, which is useful not only for selecting the optimal study design but also in helping with possible sample size calculations. For example, diseases that are rare, such as anencephaly (rare birth defect where babies are born without parts of their brain or skull), are better studied using case–control study design approaches compared to cohort study, which may not be feasible due to possibly small numbers seen at the hospital, unless specific disease registries are available.

Finally, there is some trade-off in relation to the scientific rigors of the study design and available resources, so it would help to find out from the investigator how much resource (time and money) is available for the project. If there are no resources to conduct a cohort study or manpower to collate data from the case notes for a case–control study, and if a cross-sectional study is the only available design, one could then go back to discussing the objectives again and see if a different but yet equally important alternative research question can be formulated. In the article entitled 'Deconstruction Statistical Questions' and the ensuing discussions that follow (Hand, 1994), a description of some of the important issues to be considered when attempting to deconstruct a research question is presented. Numerous examples as well as a reversal where statistical techniques are considered and then understanding what questions they can answer are also discussed.

1.5 Common Mistakes

The following are some of the common mistakes made in designing an observational study that should be avoided or considered when interpreting results from such studies:

- *Selection bias by picking subjects through non-randomised selection techniques*

  For instance, in a case–control study looking at the associated risk factors for chronic obstructive pulmonary disease (COPD) patients, investigators may be tempted to select those attending a public health education programme out of convenience. The problem is that the sample is self-selected and may include those who are predisposed to having symptomatic risk factors such as coughing and smoking. The sample is not representative of all COPD patients, and any relationship between the risk factors and COPD in that study is going to be biased.

- *Improper assessment of exposure (e.g. recall bias) in a case–control study*

  As an example, recently diagnosed lung cancer patients and controls can be asked about their history of asbestos exposure. It is likely that the lung cancer patients might be more likely to recall exposure to environmental hazards (including asbestos exposure) due to their heightened attention from their disease compared to the controls. This is known as differential misclassification, and methods including use of restricted controls to prevent recall bias have been proposed (Drews et al., 1993).

- *Selecting cases and controls based on the exposure instead of outcome variable*

  Although uncommon, when designing a retrospective case–control study, sometimes it is possible to mix up the exposure and outcome variables in the selection of the sample to include in the study. For instance, in a study looking at the association between smoking and COPD, selecting the sample based on the smoking status may not necessarily provide us with enough cases, and may cause problems in the analysis and interpretation of the data. It is also important to note that the selection of cases and controls need to be made independent of their exposure status, as this will also bias the results of the study.

- *Using a cross-sectional study to establish causality*

  A cross-sectional study that has just demonstrated a statistically significant relationship between reading statistics textbooks and depression is hardly compelling evidence that reading the textbooks causes one to have depression, or the fact that depressed people are more likely to read statistics textbooks! The main problem is that the cross-sectional study does not fulfill most of the criteria to establish causality, including temporal relationship between the risk factor and the outcome, specificity in the relationship with risk factors. If the investigator is interested in establishing causality between a risk factor and disease, then a cross-sectional study should not be recommended. If a cross-sectional study had been performed, then phrases such as 'X predicts Y' or 'X causes Y' should be avoided. Instead, 'X is associated with Y' should be used.

- *Selecting cases and controls from two different populations*

  The cases and controls should be selected from the 'same source population' to make valid inferences about the population. One common mistake in a case–control study is to have cases selected from the hospitals (e.g. through clinic records or registries), and then to compare them with controls selected from a different population (e.g. residential suburbs). The obvious problem here is that the two groups may not be comparable, especially with the latter presenting with fewer risk factors for adverse health outcomes. This is also known as the 'healthy worker' effect. However, selecting controls from the same ward may also be problematic if the risk factor is equally likely to present in both groups (e.g. cases are COPD patients and controls are other respiratory patients in the same ward that may include lung cancer patients who smoke as well). This is known as Berkson's paradox (Westreich, 2012), and may provide bias in the effect size towards the null hypothesis of no association. Therefore, controls need to be carefully selected within the same population, but in a way that does not mask the effect of the explanatory variable we want to study (e.g. different ward but same hospital).

- *Failure to specify inclusion and exclusion criteria for subjects in the study* (Clark & Mulligan, 2011)

  Specifying the inclusion and exclusion criteria, including the geographical location of the study, has implications on the generalisability and applicability of study results to other settings. For instance, a study based on indigenous people living in remote Australia in terms of risk factors for the spread of influenza will have less applicability to a country like Singapore, which is more densely populated and has a different demography and culture. One also needs to be careful that the criteria does not end up excluding people with important risk factors that is being studied (e.g. excluding mothers above 40 years in the study of risk factors for adverse birth outcomes). Having additional inclusion/exclusion criteria makes the study population more homogeneous and reduces the variability in the outcome measures. Although this may be attractive in terms of reducing the sample size required for the study, it also makes the study less generalisable, and hence, a trade-off is needed. The biostatistician should discuss all the options and implications for setting the inclusion/exclusion criteria with the collaborator during the study design deliberation process.

- *Failure to determine and report measurement error methods* (Clark & Mulligan, 2011)

  For variables that are subjective in nature (e.g. pain score or satisfaction with patient care scores), it is important that the error rate of the measurement tool is calculated and considered when evaluating the utility of such tools in the study. Both intra-rater (within subjects) variability and inter-rater (between raters) variability need to be ascertained via a small test–retest study. For continuous variables, the intra-class correlation coefficient can be used to quantify the level of agreement, whereas for categorical variables, the kappa statistic can be used. The strength of kappa agreement can be quantified using scales (e.g. 0–0.2 slight, 0.21–0.4 fair, 0.41–0.6 moderate, 0.61–0.8 substantial, 0.81–1 almost perfect; Landis & Koch, 1977). Although the cut-offs are arbitrary, the scales can be used to make relative comparisons between the various tools that are being studied. If more than one assessor is involved in the study, details of their training and protocols as well as assessment need to be considered. The implication for using a tool (e.g. pain score) that has poor inter/intra-rate agreement is quite severe. The study may not be able to detect a statistically significant difference in pain score between groups that are being studied, not because there is no true difference, but rather because there is too much variability associated with the tool. The same problem also lies with using normative values (reference range on what can be considered to be normal) for the main outcome and exposure variables, which have not been established or validated for the target population under study.

- *Failure to perform sample size calculation before a study commences* (Clark & Mulligan, 2011)

  Sample size is an important topic and is covered in more detail in Chapter 4. Performing a sample size calculation will ensure that the study is adequately powered to detect clinically meaningful differences, and this calculation needs to be performed prior to the start of the study or experiment. Some investigators replicate the sample size from related studies that they have read without realising that these studies are based on a different population with possibly different parameters and values for variables are being studied. Some may also feel that the sample size is restricted to the small number of patients that they see in their clinic, without realising the sample pool can be expanded by lengthening the period of their study, or by conducting multi-centre or multi-national studies with other collaborators. Sometimes, sample size can be reduced by opting for optimal study designs, such as a repeated measures design or a survival analysis framework. Sample size considerations should not be trivially dismissed by the collaborator at the study design stage, and the biostatistician should endeavour to explain to the collaborator the importance of embarking on a sample size/power calculation.

- *Failure to understand Simpson's paradox when planning for subgroup analysis*

When interpreting tables or subgroups in an observational study, one needs to be cautious about the possibility of Simpson's paradox, where a third factor reverses the effect first observed between an exposure and an outcome variable. Julious & Mullee (1994) reports results from another study looking at the success rates for kidney stone removal between those who underwent open surgery versus percutaneous nephrolithotomy. The overall analysis showed that percutaneous nephrolithotomy performed better than open surgery (83% vs. 78%, respectively). However, when data were stratified by the diameter of the stone removed, among those less than 2 cm, open surgery performed better than the percutaneous nephrolithotomy group (93% vs. 83%) similar to the ≥ 2 cm group (73% vs. 69%). It has been argued that the main reason why the success rate reversed is because the probability of having open surgery or percutaneous nephrolithotomy varied according to the diameter of the stones. Subgroup analysis needs to be planned carefully, and the strategy for undertaking it needs to be predetermined (e.g. through a test for interaction).

1.6 Online Tools and Resources

1. The following link to the Bio-Medical Library, University of Minnesota, provides a broad description of the various study designs, including case series and systematic reviews. http://hsl.lib.umn.edu/biomed/help/understanding-research-study-designs. Date accessed: 23 July 2013.

2. The following link and associated video from George Washington University, Washington DC, shows how one can find specific study designs and publication types in Ovid MEDLINE. http://www.gwumc.edu/library/tutorials/studydesign101/howtofind.html. Date accessed: 23 July 2013.

3. Interrupted time series analysis and useful video that explains the background, utility and software demonstration of this quasi-experimental study design that is increasingly being used to evaluate health-care and policy interventions, especially in hospital setting. https://www.youtube.com/watch?v=2CDo7B72meQ. Date accessed: 11 May 2015.

4. Practice scenarios to determine epidemiology study design and useful video that interactively goes through the various

epidemiological observational study designs. https://www.youtube.com/watch?v=SFKsNJOhbK8. Date accessed: 11 May 2015.

5. A series of useful online lectures on issues relating to observational study designs such as selection and information bias in epidemiological studies, including diagnostic studies. http://www.teachepi.org/courses/fundamentals.htm. Date accessed: 28 July 2015.

1.7 A Collaborative Case Study

Project title: A prospective cohort study on the impact of a modified Basic Military Training (mBMT) programme based on pre-enlistment fitness stratification amongst Asian military enlistees.

Collaborating clinician: Dr. Yi Ann Louis Chai, MBBS (Singapore), MRCP (London), specialist accreditation in infectious disease, PhD (Nijmegen, the Netherlands); consultant, Division of Infectious Diseases, National University Hospital Singapore, Singapore.

Outcome: Following publication. A prospective cohort study on the impact of an mBMT programme based on pre-enlistment fitness stratification among Asian military enlistees (Chai et al., 2009).

Study Summary

The aim of the study was to evaluate the effectiveness of an mBMT programme based on pre-enlistment fitness stratification, in terms of improvement in the aerobic fitness and physical performance profiles among soldiers in Singapore (Chai et al., 2009). The study design was a prospective cohort study, and the main outcome measures of interest were cardiopulmonary responses during clinical exercise testing, 2.4 km run time and body mass index. The study found that the modified group had greater improvement in cardiopulmonary indices and physical performance profiles than the direct intake group, as determined by cardiopulmonary exercise testing and 2.4 km run time.

Alternative Study Designs

Could we have used a different study design? Clearly, a cross-sectional study would not be appropriate as a temporal time period was needed between the intervention and the outcome measures to establish causality. A case–control study design would also be difficult to implement unless one dichotomises the outcome measure into 'those who improve' versus 'those who do not' (an exercise that some may be construed as arbitrary in nature). The closest alternative we considered was a randomised controlled trial, but the nature of the study involved selecting groups of recruits based on an initial screening of their medical and fitness status, which would have made it quite difficult to implement in a military hospital setting.

Key Learning Points

1. The various commonly used observational study designs in clinical research have been described in this chapter using a clinical scenario (skin cancer and solar exposure) to demonstrate the similarities and differences between the various designs, and also to show that there can be more than one way to answer the same research question.

2. The potential problems and advantages of each study design have been highlighted, so that these can be discussed with the collaborator when deliberating on the appropriate study design.

3. This chapter lists the questions to ask collaborators, especially in relation to the aims and hypothesis, which can then point towards an optimal study design that can be employed in the study.

4. Section 1.5 has discussed some common mistakes in the design of observational studies aims to equip the biostatistician with information on how to avoid selecting an incorrect design or even suboptimal features within a particular study design. Finally, the exemplar discussion on a collaborative project should highlight the rationale for making study design choices in a real-life clinical situation.

References

Armitage, P., Berry, G., & Matthews, J. (2002). *Statistical Methods in Medical Research* (4th ed.). Oxford: Blackwell Science.

Chai, L. Y., Ong, K. C., Kee, A., Earnest, A., Lim, F. C., & Wong, J. C. (2009). A prospective cohort study on the impact of a modified Basic Military Training (mBMT) programme based on pre-enlistment fitness stratification amongst Asian military enlistees. *Ann Acad Med Singapore, 38*(10), 862–868.

Clark, G. T., & Mulligan, R. (2011). Fifteen common mistakes encountered in clinical research. *J Prosthodont Res, 55*(1), 1–6. doi:10.1016/j.jpor.2010.09.002.

Drews, C., Greenland, S., & Flanders, W. D. (1993). The use of restricted controls to prevent recall bias in case-control studies of reproductive outcomes. *Ann Epidemiol, 3*(1), 86–92.

Earnest, A., Tan, S. B., & Wilder-Smith, A. (2012). Meteorological factors and El Nino Southern Oscillation are independently associated with dengue infections. *Epidemiol Infect, 140*(7), 1244–1251. doi:10.1017/S095026881100183X.

Fletcher, R., & Fletcher, S. (2005). *Clinical Epidemiology: The Essentials* (4th ed.). Philadelphia, PA: Lippincott Williams & Wilkins.

Hand, D. J. (1994). Decontructing statistical questions. *J R Stat Soc A, 157*(3), 317–356.

Hulley, S. B., Cummings, S. R., Browner, W. S., Grady, D. G., & Newman, T. B. (2007). *Designing Clinical Research*. Philadelphia, PA: Lippincott Williams & Wilkins.

Julious, S. A., & Mullee, M. A. (1994). Confounding and Simpson's paradox. *BMJ, 309*(6967), 1480–1481.

Landis, J. R., & Koch, G. G. (1977). The measurement of observer agreement for categorical data. *Biometrics, 33*(1), 159–174.

Machin, D., Campbell, M. J., & Walters, S. J. (2007). *Medical Statistics: A Textbook for the Health Sciences* (4th ed.). West Sussex: John Wiley & Sons.

Ong, M. E., Wah, W., Hsu, L. Y., Ng, Y. Y., Leong, B. S., Goh, E. S., ... & Earnest, A. (2014). Geographic factors are associated with increased risk for out-of hospital cardiac arrests and provision of bystander cardio-pulmonary resuscitation in Singapore. *Resuscitation, 85*(9), 1153–1160. doi:10.1016/j.resuscitation.2014.06.006.

Pfeiffer, R., & Riedl, R. (2015). On the use and misuse of scalar scores of confounders in design and analysis of observational studies. *Stats Med*. Advance online publication. doi:10.1002/sim.6467.

Rush, J. A. *Aims and Hypotheses—How Are They Related*. https://teamlead.duke-nus.edu.sg/vapfiles_ocs/2011/CliSci/Clinical_Research_Topics/%2828%29Aims_and_Hypotheses-How_Are_They_Related_%28Prof_John_Rush%29/. Accessed 23 July 2013.

Wah, W., Das, S., Earnest, A., Lim, L. K., Chee, C. B., Cook, A. R., ... & Hsu, L. Y. (2014). Time series analysis of demographic and temporal trends of tuberculosis in Singapore. *BMC Public Health, 14*(1), 1121. doi:10.1186/1471-2458-14-1121.

Westreich, D. (2012). Berkson's bias, selection bias, and missing data. *Epidemiology, 23*(1), 159–164. doi:10.1097/EDE.0b013e31823b6296.

2

Randomised Controlled Trials

2.1 Introduction

In Chapter 1, we looked at some common types of observational study designs. In this chapter, we examine randomised controlled trials (RCTs), a study design whereby the investigator intervenes (most commonly by introducing a new drug) and observes the effect of the intervention through measuring some outcome measures. The key advantage of an RCT over observational study design lies in the fact that it is usually easier to establish causality and confounding can be minimised. Drug trials usually employ RCTs, as they provide the highest level of evidence among all the epidemiological study designs. The various types of RCTs and key features such as randomisation and use of controls will be highlighted in this chapter. Newer trial designs, including the stepped wedge RCT, are introduced, along with a section on the role of a statistician in a Data and Safety Monitoring Board of an RCT. A unique feature of this chapter relates to the provision of issues and concerns to discuss and clarify with collaborators during the design phase of an RCT. Another important section of this chapter relates to the choice of outcome measures in an RCT and the critical role the biostatistician should play in the identification and selection of the outcomes. A real-life collaborative case study is provided at the end of the chapter.

2.2 Phases and Types of Trials

Generally, trials can be grouped into four different phases (Pocock, 2000). The first phase looks at clinical pharmacology and toxicity, with the aim of evaluating the maximum tolerable dose, as well as drug metabolism and bioavailability. Phase I trials are usually conducted in specialised units within hospitals or research institutes, as patients may need to be warded and observed for a period of time. This includes pharmacokinetic studies involving single-rising dose tolerance, multiple-dose tolerance, bioavailability, elimination half-life, food effect and bioavailability with positive control (Welling, 1997). The statistical models and techniques used to analyse phase I trials are usually quite standard and include plotting concentration response curves, area under the drug–concentration curve, absorption lag time, drug accumulation rate and use of compartment models, and these are well defined along with their mathematical derivations (Welling, 1997). More comprehensive discussion of the various statistical models available (also known as pharmacokinetic and pharmacodynamic statistics), including non-linear pharmacokinetic models, can be found here: http://www.boomer.org/c/p4/index.php?Loc=Visitor. Date accessed: 12 May 2015. This website also provides an interactive graphing tool that allows one to visualise the data. Dykstra and colleagues present a consolidated set of guiding principles for reporting of population pharmacokinetic analyses based on survey inputs as well as discussions between industry, consulting and regulatory scientists (Dykstra et al., 2015).

Phase II trials are initial clinical investigations to examine the effectiveness and safety of a drug, and are a preamble to larger phase III studies, which typically include more patients to confirm the effectiveness of the drug or intervention. Phase IV studies are also known as post-marketing surveillance studies, and are typically conducted after a drug has been approved for marketing, and used to introduce drugs to new places and settings. It is not the purpose of this chapter to discuss all aspects of a trial, but rather to discuss the specific aspects, which involve the input from the statistician in the design of the trial, including randomisation and selection of controls and performing blinding. Statistical analysis of data from trials and sample size calculation for the design of a new trial will be covered in Chapters 4 and 5.

2.3 Types and Features of Randomisation

Randomisation is the process of randomly allocating subjects to the various arms in a trial (e.g. treatment or placebo), and the investigator does not have a role in the allocation steps. In an RCT, the act of randomisation eliminates bias in allocating treatments, which is a major advantage over observational study designs. Randomisation is usually performed after patients have been registered, recruited and checked for eligibility for the trial. There are various methods of randomisation, including simple randomisation, random permuted blocks and stratified randomisation. For simple randomisation, one generates a series of random numbers and allocates patients to treatment groups, depending on the assigned values. For instance, assuming there are two groups, one generates a series of random numbers from 1 to 10 and assigns patients to group 1 when numbers 1–5 appear, and group 2 when numbers 6–10 appear.

It may be possible to end up with unbalanced numbers in each group, particularly when the sample size is small. In this instance, random permuted blocks method is a preferred option. In a simple two-block assignment, patients are assigned in a block size of two. For example, the first two patients are assigned treatment (T) and placebo (P) in sequence TP (i.e. the first patient receives T and the second patient receives P) if the digits 1–5 are drawn, and assigned sequence PT if the digits 6–10 are randomly drawn. Block sizes can be increased and changed to reduce the chances of prediction of patient assignment to treatment groups. Typically block sizes are mixed, ranging from two to six, and the block size is kept concealed by the statistician. Other methods include stratified randomisation and minimisation (Pocock, 2000). For example, in advanced stage squamous cell carcinoma, presence or absence of deep vein thrombosis (DVT) can affect the overall survival (primary outcome). Hence, the investigator may wish to balance the proportion of DVT in two arms, which is possible by considering DVT as the stratification factor.

It should be noted that stratified randomisation doesn't mean that the stratification factor will be balanced with the treatment arm. That is, there may not be 50% patients with DVT present and 50% patients with DVT absent. It just ensures that the numbers in treatment and control arms are balanced in each strata. It should also be noted that minimisation is a non-random method that ensures that treatment groups are similar in terms of several variables, and the method is particularly suitable for smaller trials, where small numbers of patients may be recruited from each site of a multi-centre trial. A detailed discussion of this method with examples is provided in this book (Altman, 1991).

Sometimes, even after randomisation, the groups may be unbalanced in terms of key confounders, but this is usually a problem in small trials, and unless there is something fundamentally wrong with the randomisation process, any difference in baseline characteristics between the two groups can be attributed to chance. It is not necessary to adjust for baseline variables in the analysis for this reason. In relation to this, one also needs to be careful about adjusting for post-randomisation variables, as these may potentially be a response to treatment (Senn & Julious, 2009). Randomisation can be performed using random number lists or increasingly computer software such as the Research Randomiser (see Section 2.9) and even Excel. Figure 2.1 shows how we can create random codes for each subject in a trial of 20 patients. One should note that because Excel refreshes the list of random numbers after each action performed in Excel, it is probably wise to copy and paste the data as values, so that the formulas are removed, and the data are subsequently rounded to the nearest whole number. One can then decide to assign the patients to the 'Treatment' group with code 1 and 'Placebo' for those with code 0 using the logical function in-built within the formula functions in Excel (Figure 2.2).

It should also be noted that the randomisation codes should be stored in a safe and secure location, and the codes should also be broken only in a medical emergency (e.g. patient suffers a serious adverse event). The randomisation list then needs to be transferred to a series of individual assignments so that the clinician or any other trial coordinator assigning the treatments to patients will not be able to predict the next assignment,

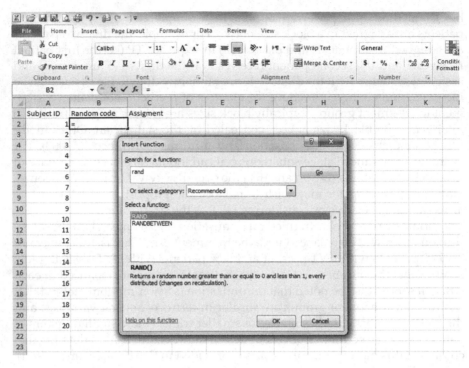

FIGURE 2.1
Creating random numbers in Microsoft Excel.

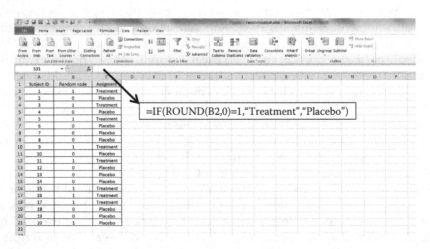

FIGURE 2.2
Using logic functions to assign groups based on the random code.

and this is often done through the use of individually sealed envelopes. This concealment of randomisation code is critical, and, if not done correctly, makes the whole randomisation process futile. The envelopes usually have the trial identifier and sequence number on the cover and the treatment allocation and patient trial number inside. Multi-centre trials can administer the randomisation list through a central office via telephone or even through web-based programmes. Academic Research Organisations such as the Singapore Clinical Research Institute provide web-based randomisation for a fee (http://www.scri.edu.sg/index.php/2012-12-10-09-47-54/our-expertise-randomisation. Date accessed: 12 May 2015).

2.4 Controls and Blinding

In an RCT, control groups are included to ensure that the treatment effects are compared. Controls are usually placebos. If there is a current standard treatment available, then it would be unethical to use placebos instead. Sometimes, there can be more than one control group (e.g. different formulations of the existing drug). Usually selection of the controls is quite straightforward, but the statistician must be careful of not comparing the intervention arm in one trial with a control arm of a separate trial. The obvious problem here is that although randomisation balances the covariates when comparing groups within a trial, it does not necessarily do so across trials. Similarly, patient profiles may differ across trials and may not be comparable. The other statistical issue lies in designing a multiple arm trial. If there are three different drug formulations to compare, then the total number of comparisons is 6, and this needs to be accounted for in the sample size calculation to address multiplicity. However, when there are two different drugs that need to be evaluated simultaneously, one may employ a factorial RCT, which enables efficient concurrent investigation of two interventions by including all participants in both analyses, as well as consider both the separate effects of each intervention and the benefits of receiving both interventions together (Montgomery et al., 2003).

Blinding is also usually performed to reduce the effect of differential reporting due to the knowledge of their treatment group and also bias in the ascertainment of the outcome from assessors or possible change in the way the clinician treats his/her patients based on the knowledge of the allocated treatment group of the patient. Blinding can be single (patient is blinded to the treatment assignment), double (both patient and investigator are blinded), or triple (patient, investigator and analyst are blinded). Pocock discusses some of the rationale and justification for blinding in a trial carefully (Pocock, 2000). In particular, there is the potential for participant response bias and outcomes assessment bias if the trial is not blinded. The benefits of blinding for patients include less biased psychological or physical responses to intervention, greater compliance with trial regimens and patients less likely to leave trial without providing outcome data, whereas for the investigators, there is less likelihood of differentially administering co-interventions, adjusting dose and differentially withdrawing participants (Schulz & Grimes, 2002).

Sometimes when interventions cannot be blinded (e.g. foot amputations among diabetics), investigators may limit and standardise other potential co-interventions as much as possible and blind staff who measure the outcomes instead (Hulley et al., 2007). Trials that are not blinded at all are called open label trials. In a completely blinded trial, the statistician analyses data with the treatment or drug group coded, so that he/she is unaware of which group of patients is assigned the drug versus placebo.

2.5 Other Types of RCTs

The following are variants of the usual RCTs one may encounter in the clinical research or population health setting.

2.5.1 Cluster RCT

Cluster randomised trials involve the randomisation of an intervention or programme to a group of individuals, rather than individuals themselves. For example, in order to evaluate the effectiveness of a health education campaign, one may wish to randomise suburbs in Eastern Melbourne into two distinct modes of health education programmes. Randomly selected participants within the selected suburbs can then be invited to participate in the programmes. Using appropriate outcomes such as incidence of disease or health risk behaviour measured at relevant timelines, one can then make the appropriate comparisons and evaluate the programmes.

The rationale for a cluster RCT is that it is often more cost-effective to administer the intervention to a group of people than to individuals themselves. In our example, providing health education talks in neighbourhood community centres may be more feasible than individually tailored talks. There is also a higher likelihood that participants within a suburb interact closely compared to across suburbs (e.g. visit the same grocery shops, exercise at the same parks, etc.) leading to possible clustering in risk factors and outcome. The major drawback of a cluster RCT is that a larger sample size would be required compared to a traditional RCT. The sample size calculation itself is more complex, with difficult decisions required around the choice of number of clusters and the inherent design effect, which in turn is affected by the intra-cluster correlation coefficient (ICC) of the key outcome variable. Campbell and colleagues provide a framework for the reporting of ICCs in cluster trials, which includes a description of the dataset and the outcome, information on the method of calculation of the ICC, information on the precision of the ICC as well as variability in ICCs (Campbell et al., 2004). For instance, it has been reported that outcomes that were measured subjectively (e.g. self-reported health status of respondents) were likely to display greater clustering than those measured objectively (e.g. laboratory-confirmed blood tests for haemoglobin).

2.5.2 Crossover Trials

Compared to the parallel trials discussed in Sections 2.3 and 2.4, crossover trials are characterised by outcomes measured for the same patients more than once. Typically, the same patients get both drugs/interventions that are being studied one after another, with the order of presentation randomised. Figure 2.3 is a hypothetical example of an RCT looking at the effectiveness of two types of corticosteroid drugs on reducing the number of exacerbations among 100 patients with chronic obstructive pulmonary disease (COPD). At baseline, patients are randomised into two groups: group 1 receiving drug A followed by drug B 6 months later and group 2 receiving drug B followed by drug A. The outcome, COPD exacerbations, is measured twice: once at 6 months of follow-up and the other at 12 months of follow-up.

An important point to note for the trial statistician is the possibility of carry-over effect, where the effect of the drug administered at baseline is still present after 6 months, when the second corticosteroid drug is prescribed. The half-life

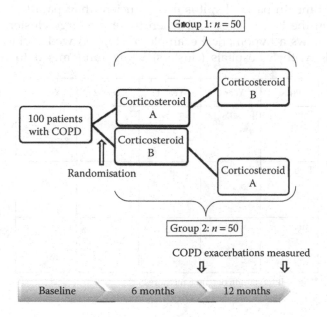

FIGURE 2.3
Typical schema for a crossover trial.

of the drug as determined through pharmacological studies as well as careful discussions with the respiratory clinician needs to be held before deciding on the follow-up period for the crossover trial. In the presence of carry-over effect, one may wish to have a suitable washout period where the patients do not receive any form of treatment. Treatment periods that are too short may mean that the treatment has too little time to take effect, whereas conversely, inadequate compliance with the protocol and substantial number of patient withdrawals can be a problem with long treatment periods (Pocock, 2000). Patients' disease status may also not remain stable over the period of the trial. The analysis of crossover trials will be described in Chapter 5.

2.5.3 Stepped Wedge Cluster RCT

The stepped wedge cluster RCT is a relatively new study design, which is an amalgamation of study features found within the cluster RCT and crossover trial designs. In such a design, all clusters (e.g. hospitals, wards, general practices, etc.) start in the control phase when there is no intervention administered. Intervention at the cluster level is then introduced in stages (steps) with randomisation used to determine which cluster should be switched over to the intervention phase. This is repeated until all clusters are eventually allocated to the intervention. Outcome measures are usually collected repeatedly over time before, during and after intervention. The stepped wedge cluster is a pragmatic design when intervention is meant to be rolled out eventually to all clusters universally, and where the outcome variables are routinely collected over time. In particular, it is more efficient than parallel cluster RCT design when the ICC is large and when there are large clusters available. Figure 2.4 shows a hypothetical example of stepped wedge cluster RCT of 15 hospitals, with 5 hospitals (clusters) being randomised to receive the

Phase	Year 0	Year 1	Year 2	Year 3
One		1–5 **	**	**
Two			6–10 **	
Three				11–15 **

** Post-intervention outcome measure

FIGURE 2.4
Schematic representation of the stepped wedge design.

intervention at each phase. There is an initial period of data collection when there is no intervention (gray shade). The intervention is a multi-disciplinary patient safety programme for pancreatic cancer patients, which was planned as a phased roll-out intervention due to logistic and administrative efficiency. The outcomes were patients' compliance to a set of quality care indicators, which were routinely measured in the hospitals.

A number of design features need to be deliberated in the planning of such a study, including number of clusters, number and length of randomised units at each step (Hemming et al., 2015). In recent times, the design has been employed in a variety of clinical settings including the studying of the effects of a theory-based continual medical education in early cancer diagnosis at three levels: general practice (GP) knowledge and attitude, GP activity and patient outcomes in Denmark (Toftegaard et al., 2014), effectiveness of a community-level physical activity intervention providing tailored physical programme to rural villages in England (Solomon et al., 2014) and optimising psychosocial recovery outcomes for people affected by mental illness in the community mental health setting in Australia (Palmer et al., 2015).

Due to the nature of the study design, samples can be taken from independent subjects (cross-sectional) or from the same patients over time (cohort). The number of measurements taken before/after intervention can vary, and sometimes, it is possible to have a transition period during intervention when there is no measurement taken. As mentioned in Sections 2.3 and 2.4, this trial design is still at its infancy stage, and there are several outstanding methodological issues that need to be resolved, including sample size calculations for cohort design as opposed to the cross-sectional nature of data collection and incorporation of the possible effects of confounding effect of calendar time. Statistical models analysing data from such trials are also not trivial and need to account for the nesting and clustering of the data as well as potential repeated measurements over time.

2.5.4 Equivalence Trials

Compared to the usual efficacy studies, equivalence trials aim to show that a new drug or a new formulation of an existing drug works just as well or effective as the existing gold standard treatment. They are usually designed when the new drug is not expected to be substantially better than the existing drug, but may offer other advantages such as lesser side effects, cheaper or an improved quality of life (QoL). Equivalence is usually demonstrated when the ratio of peak concentrations of the two drugs is between 0.8 and 1.25 (also known as the equivalence margin), as well as the entire 90% confidence interval of the ratio falling within the same range (Motulsky, 2010). One important issue for a statistician to consider in the design of an equivalence trial is the equivalence margin. Obviously, a smaller margin will require large sample sizes, but too large an effect would also have limited clinical utility.

The other statistical point to note is that compared to traditional trials, the roles of the null and alternative hypotheses are reversed in an equivalence trial. For instance, in a traditional trial looking at mean difference in response among those prescribed drug X and drug Y, the null and alternate hypotheses can be written as shown in Figure 2.5. By contrast, the null and alternative hypotheses are reversed in an equivalence trial, as shown in Figure 2.6. The null hypothesis is then written as follows: the mean difference between the two groups is $> \Delta$ (the equivalence margin); the alternate hypothesis being written as the mean difference between the two groups is $\leq \Delta$.

An important consequence here for the statistician is that if the sample size is inadequate or if the dropouts are significant, the power of the study can be reduced to reject the null hypothesis in favour of the alternative, and therefore, an inferior new treatment may appear to be equivalent to the standard when in reality it is due to an underpowered and poorly done study (Hulley et al., 2007).

Non-inferiority trials are similar trials that aim to show that a new drug is no worse than an existing one, in which case, all parts of the confidence intervals for the ratio are of the right of the lower border of the equivalence margin. A non-inferiority design is useful when ethically it may not be possible to use a placebo or 'no treatment control group', due to the existence of a current treatment which is life-saving. As shown in Figure 2.7, equivalence is demonstrated only in scenario 1, when the 90% confidence interval of the effect size ratio lies entirely within the equivalence margin. Non-inferiority, however, is demonstrated in scenarios 1 and 3, where the confidence interval of the effect size ratio does not cross the lower bound of the equivalence/non-inferiority margin. There is no universal standard on the specification of the non-inferiority margin, although this is usually specified in advance before the trial is conducted and is usually set to be half or less than the historical effect size or even a percentage (e.g. 80%) of the effect of the current treatment or what is deemed clinically unimportant (Greene et al., 2008).

A non-inferiority trial design may not be appropriate in all instances. In a recent draft guideline by the Food and Drug Administration, situations

$$H_0 : \mu_x - \mu_y = 0$$
$$H_A : \mu_x - \mu_y \neq 0$$

FIGURE 2.5
Hypothesis testing in a traditional trial that compares means in two groups.

$$H_0 : |\mu_x - \mu_y| > \Delta$$
$$H_A : |\mu_x - \mu_y| \leq \Delta$$

FIGURE 2.6
Hypothesis testing in an equivalence trial that compares means in two groups.

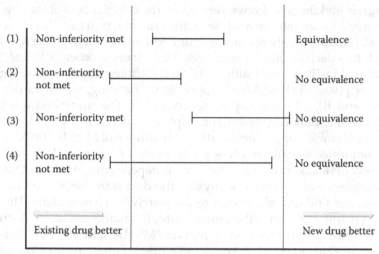

(1) Non-inferiority met   Equivalence

(2) Non-inferiority not met   No equivalence

(3) Non-inferiority met   No equivalence

(4) Non-inferiority not met   No equivalence

Existing drug better   New drug better

Equivalence margin

FIGURE 2.7
Outline of equivalence and non-inferiority studies. *Note:* Lines represent mean effect size ratios between new and existing drugs and bounded by their 90% confidence intervals.

when such a design is not warranted are outlined (US Food and Drug Administration, 2010):

1. The treatment effect may be so small that the sample size required may not be feasible.
2. There is large study-to-study variability in the treatment effect, such that it is not sufficiently reproducible to allow for the determination of a reliable estimate of non-inferiority margin.
3. There is no historical evidence to determine a non-inferiority margin.
4. Medical practice has changed so much that the effect of the active control in past studies may not be relevant to the current study.

2.6 Data and Safety Monitoring Board

Sometimes, it is possible that a statistician is not involved in the main trial, but rather in the DSMB as one of the committee members. The committee usually meets at least once while the trial is under way, and its main objective is to determine whether the trial should be terminated, modified or even extended for a longer period than originally planned (Herson, 2009). For instance, when there are significantly higher levels of serious adverse events attributed to the investigational drug, a decision may be made by the committee to prematurely end the trial. Conversely, when the effectiveness of the new drug is established via an interim analysis of the trial, the trial can also be stopped so that all remaining patients can be assigned and benefit from the new drug, although this decision has been criticised by others (Pocock & White, 1999). The National Institutes of Health (NIH) in the United States already requires DSMBs for phase III clinical trials applying for funding, but for earlier trials (phases I and II), a DSMB may be appropriate if the studies have multiple clinical sites, are blinded (masked) or employ particularly high-risk interventions or vulnerable populations (National Institutes of Health, 2000).

The committee usually receives a summary of the trial data, including adverse events and key outcomes from the independent statistician. One obvious consequence of repeatedly analysing the data is the issue of multiplicity, and thereafter finding a significant result purely by chance alone. The Food and Drug Administration in the United States has made available a document on its website (http://www.fda.gov/OHRMS/DOCKETS/98fr/01d-0489-gdl0003.pdf. Date accessed: 5 April 2015), titled 'Guidance for Clinical Trial Sponsors – Establishment and Operation of Clinical Trial Data Monitoring Committees'. Among other information, the document highlights standard operating procedures to adopt in setting up the committee, roles and responsibilities of members and also information for statisticians conducting the interim analyses (Section 6.4). In particular, it recommends that the statistician who is performing the interim analysis should not be involved in the conduct of the trial, especially in modification of the original trial design.

Several early stopping rules exist to address the issue of multiplicity. The simplest method is the Bonferroni correction, where the overall level of significance, α is divided by the number of times the analysis is performed, for example, n time, to provide a lower threshold, that is, $\alpha_i = \alpha/n$. For instance, if there are two analyses planned, each interim analysis should be performed, and results evaluated against a level of significance value of $0.05/2 = 0.025$. Disadvantages of this method include a very low level of significance for the final analysis and an unrealistic assumption that each test is independent of the other.

Other more complex sequential schemes exist, including the Pocock boundaries and O'Brien–Fleming scheme, and a discussion of the relative merits of each method is undertaken by Armitage (Armitage et al., 2002). For instance, the Pocock rule will lead to earlier termination than others if the treatment effect is very large.

2.7 Choosing Outcome Measures for a Trial

In a clinical trial, selecting the key outcome variables is crucial to the proper evaluation of the intervention. It is common to see a few key variables listed as the primary outcome variables, and for the trial to be adequately powered to detect important differences in these variables, though one needs to also be cautious about not including too many variables due to the issue of multiplicity. The outcomes are usually clinical (e.g. death and surgery) and can be disease specific (e.g. positive surgical margins for prostate cancer). Secondary and exploratory outcomes should also be listed in the protocol. All else being equal, continuous variables would be preferred over categorical ones as they retain more information and hence require a smaller sample size. Clinically important variables include mortality and incidence of disease such as cancer, acute myocardial infarction and rheumatoid arthritis, and they provide the most useful evidence in terms of treatment efficacy or effectiveness of new interventions/programmes. By contrast, questionnaire or survey-based responses such as satisfaction surveys, health scores or pain scales can be problematic due to issues regarding validity, accuracy and reliability of the instrument as well as biases in elucidating the responses from patients.

However, there are tools such as the Short Form 36, which is a generic and multi-purpose, short-form health survey with only 36 questions, yielding an eight-scale profile of functional health and well-being scores as well as psychometrically based physical and mental health summary measures and a preference-based health utility index. The questionnaire has been validated, and the reliability of the eight scales and two summary measures have been established using both internal consistency and test–retest methods.

In the absence of clinical outcomes that may take a longer time to manifest, intermediate markers that are related to the clinical outcomes can also be used. Trials that use intermediate markers can further our understanding of pathophysiology and provide useful information in designing the best dose or frequency of treatment to use for clinical outcomes in trials (Hulley et al., 2007). According to the Consolidated Standards of Reporting Trials (CONSORT) statement (Moher et al., 2012), outcomes should be completely defined with sufficient details to allow for others to replicate. If they are measured at several time points in the trial, the primary time point of interest should be prespecified, along with information about who/how many people are involved in the assessment. For instance, if QoL is measured annually for 5 years after discharge from hospital among stroke patients, the primary outcome measure may be defined as the QoL measured 1 year after discharge, and the questionnaire may be administered by two clinical research coordinators who have been trained in this area.

Sometimes, in a clinical trial, it is possible to combine several outcome measures to come up with a single composite end point. This may be done to increase the power of the study (e.g. the single end point may have very few outcomes like deaths) or even to circumvent the issue of multiplicity when analysing multiple outcomes. Sometimes, this may not work, as the individual end points may measure very different things and the results can be mixed. For instance, it has been reported in a trial among patients with myocardial infarction and left ventricular dysfunction that there was no significant difference between carvedilol and placebo when the composite end point of 'all-cause mortality or hospital admission for cardiovascular problems' was used ($P = .30$) (Senn & Julious, 2009). When the composite end point was disintegrated, it was reported that the more serious outcome of all-cause mortality was significantly lower in the carvedilol group compared with the placebo ($P = .03$).

2.8 Key Questions to Ask a Clinician

1. What is the main purpose of the trial?

   If the main objective is to establish efficacy of a new drug, then a superiority study needs to be planned. If the aim is to show that the different formulations or doses of the same drug are similar in terms of outcomes and adverse events, then a non-inferiority trial needs to be done. It would be incorrect to design a study to look at efficacy and then convert that aim to be one of non-inferiority trials when the results are not significant.

2. Can you state all the hypotheses, including specifics?

   The hypothesis will provide for the appropriate design of the trial (e.g. cluster-randomised trial to evaluate a community-based intervention). The *a priori* statement of the hypothesis will prevent any sort of post-study sieving of data and looking for significance among secondary measures or other routinely collected data. The hypothesis will also provide the metrics upon which sample size calculation can be performed.

3. Do you need controls, and if so, what is the most appropriate comparator group to use in the trial?

   Controls may consist of placebos or the current treatment regime and the choice of controls is largely determined by the specific hypothesis being tested and to a lesser extent sample size considerations. The controls may also consist of parts of the intervention. For instance, in anxiety disorder RCTs, one may test a full package of cognitive–behavioural interventions and relaxation training against relaxation or cognitive restructuring only, but these studies may require much larger sample sizes (Mohr et al., 2009).

4. Will there be any logistical/ethical issues in performing blinding in the trial?

   Some trials may be difficult to blind (e.g. surgical interventions), and for those that are possible, the mechanism, steps and other key characteristics of blinding should be provided. Sometimes, patient may also be able to infer his/her allocated treatment (i.e. unblinding) via obvious adverse events from the drug they were given or from the associated side effects.

Other aspects in the study design (such as concealing the allocation allotment) can be strengthened in the event that blinding is not possible.

5. How were the key outcome measures selected?

The outcome measures need to be able to measure changes due to the intervention, be as objective as possible and have characteristics such as good reliability, reproducibility and validity. Section 2.7 provided some guidance on choosing outcome measures for a trial.

6. Who will be generating the randomisation list, enrolling participants and assigning participants to interventions?

If the principle investigator (PI) is new to conducting an RCT, the process of randomisation, including the workflow, tasks and persons in charge of each task, needs to be explained. The statistician (or a data management centre for a multi-centre trial) usually generates and keeps a copy of the randomisation list.

7. For questionnaire-based outcomes, have these been validated and reliability established using internal consistency and test–retest methods?

The scientific validity of a trial is affected by the outcome measures used in the study. For non-objective measures such as questionnaire and psychometric measures, key domains such as validity and reliability of the instrument need to be carefully established before they are used in a specific RCT population (more in Chapter 3).

8. Do you have any preliminary data on the main outcomes, including means, standard deviations, prevalence and minimum effect size that is clinically meaningful to you?

Preliminary data are useful in establishing the feasibility, logistical issues involved in conducting a larger RCT. This is particularly useful as many grant funding authorities request for some pilot or preliminary data to accompany the main application in order to establish the competence and experience of the investigator in pursuing the research (see, e.g. the National Medical Research Council Singapore: http://www.nmrc.gov.sg/content/nmrc_internet/home/grant/compgrants/irg1.html. Date accessed: 23 June 2015). The data also provide for estimating the sample size for the main study.

9. Do you anticipate any issues in recruitment rate, potential dropouts or missing data on key variables in the trial?

If it is anticipated that there is going to be a poor recruitment rate in a single centre trial (e.g. from observing other trials conducted in the same institution), then perhaps the problem can be pre-empted by designing a multi-centre trial and boosting recruitment of patients elsewhere. If a key outcome variable is envisaged to contain excessive missing values, then an alternative outcome measure may be used or the statistical analysis plan amended to incorporate missing value imputation techniques.

10. Are there any other analysis (e.g. subgroup or additional analysis) you intend to perform?

    Subgroup analysis may require the initial sample size calculated to be inflated and the study design to be amended to ensure that there are sufficient samples from the subgroups. There are issues with performing additional analysis, specifically multiplicity (discussed in Chapter 5), and therefore, these additional analyses should be clinically justified *a priori* and the statistical methods amended accordingly.

2.9 Tools and Resources

1. This provides a link to the CONSORT Statement, which is an evidence-based, minimum set of recommendations for reporting RCTs. The 25-item checklist, along with the flow chart, provides a robust way to review how a trial is designed, analysed and reported (http://www.consort-statement.org/consort-statement/overview0/. Date accessed: 29 July 2013).

2. Here is a registry and results database of clinical trials from around the world (http://clinicaltrials.gov/. Date accessed: 29 July 2013).

3. The Cochrane Central Register of Controlled Trials is a bibliographic database that provides a highly concentrated source of reports of RCTs, sourced from MEDLINE and EMBASE, and records retrieved through hand searching (planned manual searching of a journal or conference proceedings to identify all reports of RCTs and controlled clinical trials) (http://community.cochrane.org/editorial-and-publishing-policy-resource/cochrane-central-register-controlled-trials-central. Date accessed: 7 July 2015).

4. Research Randomiser. Free online tool that can generate random numbers and allocations (http://www.randomizer.org/. Date accessed: 17 November 2014).

2.10 A Collaborative Case Study

Project title: Randomised controlled trial of nutritional supplementation in patients with newly diagnosed tuberculosis and wasting.

Collaborating clinician: Professor Nick Paton, MB BChir (Cambridge), MRCP (Int Med) (London), MD (Cambridge), FRCP (Edinburgh), DTM & H (UK); research director, University Medicine Cluster, National University Hospital, Singapore.

Output: Following publication. An RCT of nutritional supplementation in patients with newly diagnosed tuberculosis and wasting (Paton et al., 2004)

Study Summary

The effects of early nutritional intervention on lean mass and physical function in patients with tuberculosis and wasting was assessed via an RCT as described below (Paton et al., 2004). After completing the baseline assessments, patients were randomly assigned to either the nutritional supplement group or the control group. The randomisation was 1:1 for the two groups and was performed by randomly shuffling opaque envelopes containing study codes. Preparation of the randomisation envelopes was performed by a member of the staff who was not directly involved in the study. The study found that nutritional counselling to increase energy intake combined with provision of supplements, when started during the initial phase of tuberculosis treatment, produced a significant increase in body weight, total lean mass and physical function after 6 weeks.

Alternate Method of Randomisation

Another way one could have performed random allocation is via random permuted block size of four. As the sample size in the trial was relatively small, it would have helped in resulting in a better balance in numbers between the two groups ($n = 19$ and 17 in the intervention and control groups, respectively).

Significance of Trial

It is the first known RCT of nutritional supplementation in patients with tuberculosis.

Rigorous and Appropriate Analysis

The study made use of appropriate statistical techniques, such as the analysis of covariance model to adjust for imbalances at baseline, as well as the Bonferroni correction technique to adjust for multiplicity.

Key Learning Points

1. Understand the important features of RCTs such as randomisation, concealment of randomisation codes, use of controls and blinding.
2. Recognise other types of RCTs such as cluster RCT, stepped wedge cluster RCT and equivalence trials.
3. Define the roles and responsibilities of a biostatistician in a DSMB.
4. Identify the key issues and pitfalls in the selection of outcome measures for an RCT.
5. Learn about the 10 key questions you should ask your collaborator when designing and analyzing an RCT.

References

Altman, D. G. (1991). *Practical Statistics for Medical Research*. London: Chapman and Hall/CRC Press.

Armitage, P., Berry, G., & Matthews, J. (2002). *Statistical Methods in Medical Research* (4th ed.). Oxford: Blackwell Science.

Campbell, M. K., Grimshaw, J. M., & Elbourne, D. R. (2004). Intracluster correlation coefficients in cluster randomized trials: Empirical insights into how should they be reported. *BMC Med Res Methodol, 4*, 9. doi:10.1186/1471-2288-4-9.

Dykstra, K., Mehrotra, N., Tornoe, C. W., Kastrissios, H., Patel, B., Al-Huniti, N., ... & Byon, W. (2015). Reporting guidelines for population pharmacokinetic analyses. *J Pharmacokinet Pharmacodyn*. doi:10.1007/s10928-015-9417-1.

Greene, C. J., Morland, L. A., Durkalski, V. L., & Frueh, B. C. (2008). Noninferiority and equivalence designs: Issues and implications for mental health research. *J Trauma Stress, 21*(5), 433–439. doi:10.1002/jts.20367.

Hemming, K., Haines, T. P., Chilton, P. J., Girling, A. J., & Lilford, R. J. (2015). The stepped wedge cluster randomised trial: Rationale, design, analysis, and reporting. *BMJ, 350*, h391. doi:10.1136/bmj.h391.

Herson, J. (2009). *Data and Safety Monitoring Committees in Clinical Trials*. New York: Chapman & Hall.

Hulley, S. B., Cummings, S. R., Browner, W. S., Grady, D. G., & Newman, T. B. (2007). *Designing Clinical Research*. Philadelphia, PA: Lippincott Williams & Wilkins.

Moher, D., Hopewell, S., Schulz, K. F., Montori, V., Gotzsche, P. C., Devereaux, P. J., ... & Altman, D. G. (2012). CONSORT 2010 explanation and elaboration: Updated guidelines for reporting parallel group randomised trials. *Int J Surg, 10*(1), 28–55. doi:10.1016/j.ijsu.2011.10.001. S1743-9191(11)00565-6 [pii].

Mohr, D. C., Spring, B., Freedland, K. E., Beckner, V., Arean, P., Hollon, S. D., ... & Kaplan, R. (2009). The selection and design of control conditions for randomized controlled trials of psychological interventions. *Psychother Psychosom, 78*(5), 275–284. doi:10.1159/000228248.

Montgomery, A. A., Peters, T. J., & Little, P. (2003). Design, analysis and presentation of factorial randomised controlled trials. *BMC Med Res Methodol, 3*, 26. doi:10.1186/1471-2288-3-26.

Motulsky, H. (2010). *Intuitive Biostatistics. A Nonmathematical Guide to Statistical Thinking* (2nd ed.). New York: Oxford University Press.

National Institutes of Health (2000). *Further Guidance on a Data and Safety Monitoring for Phase I and Phase II Trials*. http://grants.nih.gov/grants/guide/notice-files/NOT-OD-00-038.html. Date accessed: 20 November 2014.

Palmer, V. J., Chondros, P., Piper, D., Callander, R., Weavell, W., Godbee, K., ... & Gunn, J. (2015). The CORE study protocol: A stepped wedge cluster randomised controlled trial to test a co-design technique to optimise psychosocial recovery outcomes for people affected by mental illness in the community mental health setting. *BMJ Open, 5*(3), e006688. doi:10.1136/bmjopen-2014-006688.

Paton, N. I., Chua, Y. K., Earnest, A., & Chee, C. B. (2004). Randomized controlled trial of nutritional supplementation in patients with newly diagnosed tuberculosis and wasting. *Am J Clin Nutr, 80*(2), 460–465.

Pocock, S., & White, I. (1999). Trials stopped early: Too good to be true? *Lancet, 353*(9157), 943–944. doi:10.1016/S0140-6736(98)00379-1.

Pocock, S. J. (2000). *Clinical Trials: A Practical Approach*. Chichester, U.K.: John Wiley & Sons.

Schulz, K. F., & Grimes, D. A. (2002). Blinding in randomised trials: Hiding who got what. *Lancet, 359*(9307), 696–700. doi:10.1016/S0140-6736(02)07816-9.

Senn, S., & Julious, S. (2009). Measurement in clinical trials: A neglected issue for statisticians? *Stat Med, 28*(26), 3189–3209. doi:10.1002/sim.3603.

Solomon, E., Rees, T., Ukoumunne, O. C., Metcalf, B., & Hillsdon, M. (2014). The Devon Active Villages Evaluation (DAVE) trial of a community-level physical activity intervention in rural south-west England: A stepped wedge cluster randomised controlled trial. *Int J Behav Nutr Phys Act, 11*, 94. doi:10.1186/s12966-014-0094-z.

Toftegaard, B. S., Bro, F., & Vedsted, P. (2014). A geographical cluster randomised stepped wedge study of continuing medical education and cancer diagnosis in general practice. *Implement Sci, 9*, 159. doi:10.1186/s13012-014-0159-z.

US Food and Drug Administration (2010). *Guidance for Industry Non-Inferiority Clinical Trials*. www.fda.gov/downloads/Drugs/.../Guidances/UCM202140.pdf. Date accessed: 19 November 2014.

Welling, P. G. (1997). *Pharmacokinetics* (2nd ed.). Washington, DC: American Chemical Society (ACS).

3

Form Design and Database Management

3.1 Introduction

Creation of an efficient and appropriate data collection form and database is an important component of the research process, and where possible, the collaborating biostatistician should be part of this process. How we design the questions/instruments will ultimately influence the kind of data that are keyed in, and ultimately analysed and interpreted. As they say 'garbage in, garbage out', and it is often very difficult to fix the problem once the data have been collected and keyed into the dataset. For instance, the problem of data entry errors is quite prevalent. Clark and Mulligan (2011) have reported on a study that evaluated the frequency and characteristics of data entry errors in large clinical databases and found that error rates ranged between 2.3% and 26.9%, with the errors being not just mistakes in data entry, but many non-random clusters in errors that could potentially affect the outcome of the studies.

Suppose one is interested in identifying the relationship between age and risk of cardiac arrests, and wants to establish a cut-off point for age when the increase in risk is most marked. This may be problematic if age is recorded as a categorical variable (e.g. <30, 31–40, >40 years) in the database even though the actual age of the respondents was collected in the questionnaire. The efficient design of a database will ensure that there are no potential errors or delays in the analysis of the study due to rectifying errors in the database. It is worth mentioning that real-life datasets are very different from those that are traditionally used in schools and textbook examples, in that the former often include 'non-sterile' datasets with incorrectly coded, labelled variables and punctuated with missing values, which could have been avoided by the proper design and evaluation of the data collection system.

This chapter begins with the best practices in the design of questions in a data collection form/questionnaire, followed by the key elements in the validation and translation of questionnaires. An introduction of types of variables and the scale of measurement is then provided as this has an impact on the type of statistical tests that can be subsequently performed. Following the data collection form design, Section 3.4 describes suitable databases that can be created for a research project. In particular, Microsoft Excel is highlighted as this is the most commonly available data entry tool utilised in the hospital setting. Tips and tricks within Excel, as well as a section on the problems of using Excel as a research database, are highlighted. Section 3.6 discusses some key questions to ask a clinician when designing a database or spreadsheet meant to pre-empt many of the problems that follow from the inefficient design of a database and also to ensure that the clinician is closely involved in the data collection phase of the study. Finally, another unique contribution in this chapter is as follows: 'common mistakes to avoid in questionnaire and database design' are presented so that these can be imparted to study personnel who are involved in the data collection and data entry stages, and help ensure that these issues do not cause problems downstream in the data analysis and interpretation stage of the project.

3.2 Principles of Questionnaire Design

Very often, data in clinical research are collected through structured questionnaires, and a summary of good procedures to adopt in the design of a questionnaire is provided by Hulley et al. (2007) and described in Section 3.3. In clinical trials, these questionnaires are often called case report forms (CRFs), and these can be in both paper and electronic forms. The CRFs should be standardized to address the needs of all those who handle the data such as the principal investigator, data manager, biostatistician, clinical research coordinator and data entry personnel. Bellary and coworkers provide details on methods of CRF design and discusses the challenges encountered and measures to be taken to prevent the occurrence of issues in a recent article (Bellary et al., 2014). The CRFs or questionnaires should not be unduly long and tedious to answer, and should include only questions which are going to be directly analysed and useful in addressing the hypotheses.

3.2.1 Open-Ended versus Closed-Ended Questions

Wherever possible, questions should be provided with pre-coded responses (closed-ended) as these are easier to answer and analyse. Sometimes, they may not be exhaustive, and options such as 'Others, pls specify' can be used. Responses should be mutually exclusive to ensure clarity. For instance, someone aged 40 years old may find it difficult to select the appropriate response for a question of age with the following pre-codes: <30, 30–40, 40 years and above!

3.2.2 Multiple-Response Questions

For questions that allow for multiple responses, such as pre-existing comorbidities, it is better to provide the list of comorbidities, and the list of pre-coded responses such as 'yes', 'no' and 'not sure'. This can help us differentiate the genuine missing responses rather than treating them as 'no's by default.

3.2.3 Double-Barrelled Questions

One should avoid double-barrelled questions that contain more than one concept or point in a single question, such as 'Do you smoke or drink alcohol regularly?'. The respondent may not be clear on how to answer the question, and any response we get may also not be useful in the analysis. Sometimes, it is possible that the investigator may wish to study the risk factor as a whole (i.e. those who drink or smoke as a group), but this can easily be coded at the analysis stage and it is good practice to collect data in the most raw or fine form at the data collection stage.

3.2.4 Wording of Questions

The wording of the questions will affect the reproducibility of the results, and the language needs to be kept clear, simple and non-technical, wherever possible. For instance, in a study of no-show at outpatient clinics, missed appointments can be used instead. The previous question on smoking can be refined as 'Do you smoke more than 10 cigarettes a day in a typical week?'. It is also a good idea to pre-test the questions on a few subjects before rolling it out for the main study.

3.2.5 Ordinal Scales

Sometimes, it may be necessary to elicit responses to a question on an ordinal scale, such as a Likert scale. For example, a patient's current quality of life (QoL) can be assessed using the following five-point scale: very healthy – 1, healthy – 2, neutral – 3, unhealthy – 4, very unhealthy – 5. Multiple questions may also be needed. Respondents may then find it relatively easier to answer the questions with such a scale rather than a simple yes/no pre-code. Statistical models such as the ordinal logistic regression can be used to analyse the data for a single question. A number of different questions can be summed up over possible domains (e.g. mental and physical health), allowing one to calculate average scores.

3.2.6 Validation

When designing a questionnaire, especially for the first time, it is important to validate and accurately translate (wherever necessary) the items in the questionnaire. Validation involves a number of steps, including establishing reproducibility (the extent to which the instrument or question provides the same results when repeated among the same subjects with no change to their health status), responsiveness (the ability of the instrument to measure changes to the subject's health status or outcomes between people with and without disease or over two time points before/after acquiring disease), validity (the extent to which the instrument is able to measure the concept it is designed to study) and reliability (the ability of various raters/assessors to provide for the same results when administering the instrument).

The total variability in the response obtained from a scale in a questionnaire includes both systematic variation between subjects and measurement error, and hence the ratio of variability between patients to the total variability (sum of patient variability and measurement error) will provide a number ranging from 0 (no reliability) to 1 (perfect reliability) (Streiner & Norman, 1995). The following equation demonstrates the mathematical formulation of one such reliability index:

$$\text{Reliability} = \frac{\sigma_s^2}{\sigma_s^2 + \sigma_e^2}$$

(3.1)

where:
   σ_s^2 denotes the between subject variability
   σ_e^2 represents the variability due to measurement error

Commonly used statistics to establish the reliability of a tool include the intra-class correlation coefficient, Cohen's kappa and the Bland–Altman plots (Bland & Altman, 1986). In the event of poor reliability, measurement error can be reduced by providing training for observers and having standardised protocols. For scales that have a ceiling effect (i.e. most responses falling within the maximum value possible) or a floor effect (most falling within the minimum value) and hence resulting in low overall variability, the scales can be modified and shifted accordingly. In Chapter 4 of the *Health Measurement Scales* book, a number of other suggestions to increase the reliability of questionnaires are also provided (Streiner & Norman, 1995). Validity is typically classified into three different kinds: content validity, criterion validity and construct validity; reliability places an upper limit on validity, so that the higher the reliability, the higher the maximum possible validity, except for the relationship between internal consistency (usually related to a number of questions measuring the same trait or characteristic, and part of reliability measure and typically analysed with the Cronbach's alpha statistic) and content validity (Streiner & Norman, 1995). Someone in the collaborative team (PI or others) may be experienced in the psychometric evaluation of the questionnaire, but the statistician usually advises on the sample size calculation and study design for establishing the reliability statistics as well as the interpretation of the values. Although it is easier to make comparisons of reliability values across tools (e.g. 0.90 is higher than a tool that provides a value of 0.7), how much is considered 'good' is a difficult call to make. One reference has suggested a value of 0.85 (Weiner & Steward, 1984), but this value is rather arbitrary and not universally accepted.

3.2.7 Translation

The translation process would usually involve two to three forward and backward translations by native speakers of both languages. It is not the aim of this chapter to discuss these concepts in detail, but the Institute for Health and Care Research provides details and checklists of the processes involved in the translation of questionnaires and forms for research. (http://www.emgo.nl/kc/preparation/research%20design/8%20 Questionnaires%20selecting,%20translating%20and%20validating.html. Date accessed: 26 May 2015). The article by Sperber (2004) also provides useful tips on the methodological problems associated with translating questionnaires for use in cross-cultural research. If the questions have been created in Microsoft Word, the readability of the questions can also be ascertained using the Flesch Reading Ease test, which uses the average number of syllables per word and words per sentence to establish a readability test scale (https://support.office.com/en-us/article/Test-your-documents-readability-0adc0e9a-b3fb-4bde-85f4-c9e88926c6aa. Date accessed: 26 May 2015).

3.3 Types of Variables and Scales of Measurement

Before setting up a database for a study, it is important to understand the different types of variables so that data can be keyed in the appropriate format. Broadly, the variables in a study can be classified into the following categories:

Outcome variables. These are also known as dependent variables and are usually the key variables of interest in the study. Examples include mortality, length of hospital stay and QoL.

Predictor variables. These are sometimes called independent variables or explanatory factors. Examples include demographic data such as age, gender, treatment groups and interventions such as new drug assignment or a novel treatment/surgery.

Confounding variables. These are variables that are associated with the predictor variable and could possibly affect the outcome variable (Hulley et al., 2007). Confounding is especially of concern in an observational study design and needs to be carefully studied, and appropriate variables measured and included in the database. Examples include presence of comorbidities and age.

Depending on the aims and context of the study, some variables can be classified into multiple categories. For example, in a cluster randomised controlled trial in the community looking at the effects of physical activity on QoL, one might use the Short-Form (SF) 36 to examine both physical and mental domains of QoL as an outcome measure. However, if the aim of the study was to examine how QoL affects mortality and the subjects are followed up for mortality, then SF 36 can be classified as an explanatory variable. However, if the researcher is interested in the hypothesis that greater physical activity will lead to a reduced mortality, then QoL is a possible confounder, and hence, SF 36 can be classified as a confounding variable.

Variables can also be classified according to the following types (Hulley et al., 2007):

Continuous. These are variables that can be quantified on an infinite scale and only limited by the sensitivity of the machine measuring them. Examples include systolic and diastolic blood pressures and various laboratory test parameters such as low-density lipoprotein and creatinine. Some continuous variables are limited to integers or counts and are termed discrete. Medical examples include the number of falls in a ward, methicillin-resistant *Staphylococcus aureus* (MRSA) infections, and so on. Count data may require a different form of statistical modelling (e.g. Poisson regression) compared to non-count continuous data (e.g. ordinary least squares linear regression). Sometimes, continuous variables are also characterised as interval compared to ratio measurements. The former is characterised by not having a true zero, and hence, ratios are not allowed (e.g. temperature in degree Celsius where 20°C is not twice as hot as 10°C).

Categorical. Variables that are not continuous can be grouped into categories. These include dichotomous (two categories) and more than two categories. The variables can be further classified as nominal (i.e. categories do not have any inherent ordering) or ordinal (there is an in-built structure and order in the way the categories are defined). Examples of nominal variables include gender (male/female), race (Indian/Malay/Chinese) and ordinal scales (rating score 1–10, where 1 indicates extremely satisfied and 10 extremely dissatisfied). Sometimes, continuous variables such as length of stay can be categorised (e.g. long-stayers vs. non-long-stayers). It should be noted that information can be lost when data are categorised, and therefore, this should only be done when there are clinical justifications (e.g. identifying prognostic factors of long-stayers).

Dates. Another common format is dates, usually in the DD/MM/YYYY format. By themselves, dates are not usually analysed. However, derived variables from dates are created commonly in databases or in the analysis. Examples include the year of admission (cohort), calculating age at admission from the date of admission and the date of birth fields, survival time from the date of admission to the date of death, and so on. It is important to state clear instructions on how to record dates and to use a consistent format throughout the data collection form or dataset.

Other forms of data. Sometimes, data are presented in rates or ratios (e.g. 20 birth defects per 1000 births in a particular suburb, or standardised incidence ratio of death, which can be defined as observed deaths divided by the expected number of deaths in town). Sometimes, data can be censored (particularly left censoring). For instance, in certain laboratory tests, there may be a threshold value, below which the machine or technology is unable to measure the parameter (non-detectable levels). A decision needs to be made about the value to input in these instances (e.g. two-thirds the detectable level or the actual detectable level), but once this decision is made, the values need to be keyed in consistently in the database.

For each of the variables in the study, operational definitions need to be provided, along with the unit of measurement. For instance, cholesterol can be measured in milligrams per decilitre of blood (mg/dL) from a blood sample collected 12 h after fasting.

3.4 Finding a Suitable Database Software for Your Study

There are various statistical software packages that allow one to key in the dataset into spreadsheet formats, such as SPSS, Stata, SAS and R. Some allow for the dataset to be imported into the software from a text or other similar formats. The most common form of spreadsheet used to key in data as a research database is Excel. Typically, after the database and study is completed, the Excel data are then converted into the statistical software that the biostatistician collaborator is most comfortable in using it. Normally, the transfer is quite easy, and there are specialised software packages, such as Stat/Transfer (http://www.stattransfer.com/. Date accessed: 1 December 2015), that allow for the seamless transfer of datasets between various software packages while preserving the underlying codes and data labels. Other more complex databases exist (e.g. Microsoft Access, REDCap, Oracle Clinical), but these are usually relatively more difficult to set up and are costly in some instances. These databases do possess some valuable features: (1) they are secure – unique login details allow only study personnel to access the database; (2) they are transparent – any changes made are traceable; (3) they have useful tools such as tracking progress of study, data query generation and study calendar and (4) they enable data import and extraction that is compatible with most statistical software. However, the Excel database is most suitable for clinical research projects based at the hospital and is the most commonly available database.

3.5 Efficient Ways to Create and Manage an Excel Database

An Excel spreadsheet is typically characterised by rows and columns. Each row represents a patient or subject, and the columns are the variables in the dataset, although for repeated measurement studies, variations exist. Figure 3.1 exemplifies a typical dataset created in Excel. The rows represent data from each patient, and the columns represent the variables. The first row shows the variable names. The variable name should be typically of 10 characters or less and should not start with number or have space between words. It is helpful to have variable names in lowercase so that there is consistency and efficiency when writing programs in statistical software. Variable names should also be unique, and in the case of multiple measurements for each patient, it is fine to attach numbers at the end of the variable name to represent the various time points when the measurement was taken (e.g. QoL1, QoL2 and QoL3).

	A	B	C	D	E	F	G	H	I	J
1	Pat ID	Sex	Age	Smoking	Heart Dis	Mortality	QolL1	QoL2	QoL 3	
2	1	0	36	1	0	1	77	49	11	
3	2	1	84	0	1	0	45	93	33	
4	3	0	80	0	0	1	66	2	76	
5	4	0	44	0	0	0	97	17	14	
6	5	0	88	1	0	0	3	46	81	
7	6	1	97	1	0	1	54	12	22	
8	7	1	76	0	1	0	6	52	4	
9	8	0	80	0	0	1	62	63	98	
10	9	0	51	1	1	0	17	1	34	
11	10	0	26	1	1	0	97	93	2	
12										

FIGURE 3.1
Example of an Excel spreadsheet.

The data in a spreadsheet should preferably be keyed in a numeric format. For categorical variables, pre-codes can be assigned for each category within the variable (e.g. 1 for males and 0 for females), and there should be a separate data dictionary sheet listing the data labels and values as shown in Figure 3.2.

3.5.1 Multiple-Response Questions

It is sometimes possible for respondents to tick multiple responses for the same question. For example, they may report different modes of exercise in a typical month. The correct strategy in creating a database is to split each of the categories as dummy variables in separate columns (Figure 3.3). The database on the left does not make it conducive at all to analyse as many different combinations of values are created, and statistical software packages

	A	B	C
1	Variable name	Variable description	Codes
2	Pat ID	Patient Identifier	
3	Sex	Gender of patient	1: male 0: female
4	Age	Age at admission (in yrs)	
5	Smoking	Smoker?	1: yes 0: no
6	Heart Dis	Presence of heart disease	1: yes 0: no
7	Mortality	Died during admission	1: yes 0: no
8	QoIL1	Quality of life at admission	
9	QoL2	Quality of life at 1 month	
10	QoL3	Quality of life at 2 month	

FIGURE 3.2
Example of an Excel spreadsheet data dictionary.

	A	B	C	D	E	F	G	H
1	Incorrect			Correct				
2	Pat ID	Exercise		Pat ID	walking	jogging	swimming	cycling
3	1	Nil		1	0	0	0	0
4	2	Jogging and swimming		2	0	1	1	0
5	3	Swimming		3	0	0	1	0
6	4	Swimming/Jogging/Cycling		4	0	1	1	1
7	5	1,2,3		5	1	1	1	0
8	6	1/2/3		6	1	1	1	0
9	7	2-3		7	1	1	1	0
10	8	1		8	1	0	0	0
11	9	1,3		9	1	0	1	0
12	10	3		10	0	0	1	0
13								

FIGURE 3.3
Keying in a multiple response question.

usually treat each combination as a unique category, which is not ideal for most epidemiological studies.

3.5.2 Repeated Measurements

Similarly, in clinical studies, measurements can be repeated for the same patient or subject. For instance, in our example in Section 3.3, we looked at QoL for each patient measured at three different time points (at admission, follow-ups at months 1 and 2). In particular, repeated measures are common in longitudinal and cohort studies as well as in hospital-based administrative records, which often include repeated visits from patients. There are two different ways to key in such data. The first is the wide format in which each measurement is keyed into a separate column (Figure 3.4). This can make the database look quite cumbersome if there are too many repeated measures or variables with repeated measures. An alternative method is to key the data in a long format in which there is one column for the measurement (QoL) and have another new variable (visit) representing the different time points the measurement was taken. Note now that each row does not represent a unique patient, but the combination of patient ID and visit would be unique. In fact, this format is most suitable for statistical software packages if one is keen to undertake complex modelling such as random effects modelling and generalised estimating equations.

3.5.3 Missing Data

If there are any missing observations for any of the variables, it is often best practice to leave the cells empty. Some software packages, like SPSS, allow for the coding of missing data (e.g. 99 or −1), but it is essential to make this assignment correctly. Otherwise, there can be potential for errors in the analysis. Sometimes,

	A	B	C	D	E	F	G	H
1		Wide format					Long format	
2	Pat ID	QoIL1	QoL2	QoL 3		Pat ID	Visit	QoL
3	1	77	49	11		1	1	77
4	2	45	93	33		1	2	49
5	3	66	2	76		1	3	11
6	4	97	17	14		2	1	45
7	5	3	46	81		2	2	93
8	6	54	12	22		2	3	33
9	7	6	52	4		3	1	66
10	8	62	63	98		3	2	2
11	9	17	1	34		3	3	76
12	10	97	93	2				

FIGURE 3.4
Keying in multiple measurements for the same patient.

in the analysis, one may wish to study those missing observations as a category, in which case it is necessary to code them with a separate number.

3.5.4 Data Validation in Excel

Microsoft Excel has some useful field validation tools that may help to facilitate data entry and minimise errors. For example, in the database example in Section 3.5.1, one might create a drop-down menu for gender, allowing for the selection of just '1's and '0's. Firstly, select the entire B column, which you intend to apply the validation rule to. Click on 'Data', and under Data Tools, select 'Data Validation'. In the Data Validation window that pops up, select 'List' under Allow, and specify '0' and '1', separated by a comma under Source. Then click 'OK'. Figure 3.5 shows the steps involved. When you now click on any cell under column B, Gender, a drop-down menu will appear, only allowing the data entry person to select 0 or 1. Alternatively, one might also specify Whole Number under Allow, and key in 0 and 1 under Minimum and Maximum, respectively. With this option, a drop-down menu does not appear, but when one attempts to key in anything other than 0 or 1, there will be an error message. This is particularly useful for validating variables such as age, so that a data entry person does

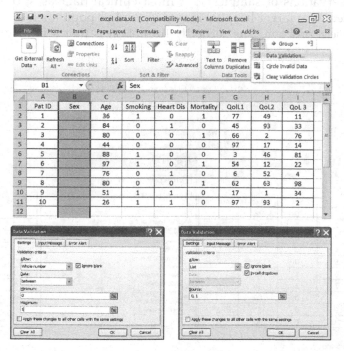

FIGURE 3.5
Data validation through creating a drop-down list.

not accidentally key in 533 instead of 53, which can give rise to disastrous results in the study.

As a rule of thumb, it is worth checking a random sample of 5%–10% of the records. Attention is needed on questions that have missing response, and their corresponding questionnaires should be checked. Logic checks such as the date of birth recorded appearing after the date of diagnosis should be performed. After all the data have been keyed in and checked, it is a good practice to lock the datasheet so that further changes cannot be made. Figure 3.6 shows us how to do this. Click on the Review tab, and then 'Protect Sheet'. Make sure 'Protect worksheet' is checked, and then type in a password that would subsequently allow user to make any changes to the sheet. Finally click 'OK', and your dataset is now password protected. Besides password protecting the dataset, it is also a good practice to back up the dataset regularly and archive the backed up dataset, in case the analysis needs to be revisited at a later stage.

3.5.5 Limitations to Excel

There are three main limitations to using Excel as a research database. Firstly, Excel has restrictions on the amount of data that can be keyed in. For example, Excel 2003 has a maximum of 256 columns (variables) and 65,536 rows (observations or samples). For most clinical databases, this should be sufficient. For larger hospital records based or population studies, this might pose a problem. The secondly limitation is that it is easy to make

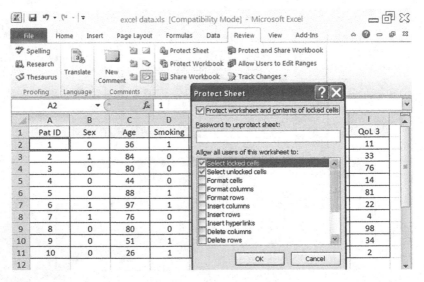

FIGURE 3.6
Locking a datasheet.

mistakes in data entry or even accidentally delete observations. In some ways, this problem can be mitigated by implementing the field validation tools and locking the datasheet as described in Section 3.3. Thirdly, Excel is not useful if the study is multi-centre in nature, if randomisation across institutions is involved or data need to be extracted and merged from multiple sources (e.g. pharmacy records with inpatient hospital admissions, etc.). If data are extracted from several sources, it can be difficult to sort data based on common identifiers and merge the datasets. In that case, web-based research data software, such as Oracle Clinical (http://www.oracle.com/us/products/applications/health-sciences/e-clinical/clinical/index.html. Date accessed: 1 December 2015) or Phase Forward Clintrial (http://www.oracle.com/us/corporate/Acquisitions/phaseforward/index.html. Date accessed: 1 December 2015), may be more appropriate.

3.6 Key Questions to Ask a Clinician

1. Is your dataset organised in a single spreadsheet in an individual 'case by variable' format?

   Sometimes, clinicians like to keep data in multiple sheets within the same Excel file or in separate Excel files due to the nature of data collection (e.g. demographic and laboratory data from different sources, separate sheets to collect baseline and follow-up values of the same parameters, etc.). As much as possible, advise the clinician to collect all data in a single Excel 'sheet' (remember not 'file'). If data are spread across different sheets or files, it is possible to merge them using statistical software tools, such as the Stata 'merge' command.

2. How many variables are there in your study?

   A large number of variables may have been collected, but if only a small subset of variables is going to be used in the study, then identifying and sub-setting the variable list may help in the analysis, especially in ease of formatting the variables, undertaking quality control checks and writing commands for the analysis. For complex statistical analysis, the speed of computation can also be improved by storing less observations. Streamlining the variables at the database formulation stage also helps to prevent the collaborator from engaging in any possible data dredging exercise.

3. What are the outcome, explanatory and confounding variables (if any)?

   Asking the clinician to identify the key variables that will be used in the analysis/study helps to eliminate the unnecessary variables from the dataset.

4. On what scale is each of these variables measured?

The scale of measurement of each variable in the study is important in determining the statistical analysis plan. For instance, if a pain score is measured on a continuous scale, then a linear regression would be appropriate, or if it is determined to be ordinal, then perhaps an ordered logistic regression can be used. At the data collection phase, this will also inform how the data are to be keyed in.

5. Is any of the variables multiple response in nature?

Questions that elicit multiple responses need careful consideration in the database design stage. They need to be keyed into separate columns (e.g. type of comorbidity: Asthma, COPD, etc.). Subsequent variables can be computed, for instance calculation of the total number of comorbidities.

6. Are you planning to have follow-ups in your study, and what are the variables that are going to be measured?

For longitudinal studies in particular, sometimes the same measurement tool is applied several times to the patient (e.g. QoL at hospital admission, 3- and 6-month follow-ups), so the variable name needs to be carefully defined to reflect this (e.g. qol1, qol2, qol3). Clinicians like to colour code follow-up data to differentiate between them, but they should be advised not to, as statistical software usually do not recognise colour codes, and this will result in unnecessary work to rectify the data.

7. Is there any major source of bias in recording any of the variables in the study?

Some variables such as body mass index may not be routinely collected in clinical practice, and if there is any systematic bias in the missingness of the data, then much effort is needed to complete data collection. If there is subjectivity in the data (e.g. patient self-reported weight readings or weight recorded at different follow-up periods for different patients), then such data may have limited utility in the analysis, and possible alternative variables should be discussed with the clinician, or the initial hypothesis should be revised.

8. How are dates (if any) formatted in your database?

It is crucial that dates are properly formatted (i.e. recognised as dates) as well as standardised for all patients. Mixing up the months and days can have consequences in the analysis, and in some statistical software packages, incorrectly imputed dates may not get recognised and hence coded incorrectly as missing observations.

9. Have you aggregated or computed any variables in your dataset?

As much as possible, data should be keyed into the database in their original form, and any new derived variable should be created using the statistical software. For instance, if one is interested in time from hospital admission to initiation of treatment, then the dates of admission and treatment should be keyed into the database, and a derived variable 'time from hospital admission to initiation of treatment' created and labelled in the dataset.

10. Are you or your research assistant familiar with using Excel or other software as a database management tool?

    It is a good idea to gauge the data management software proficiency of the collaborators before asking them to create the database. Those who are new can be pointed to resources, including this chapter that can help provide useful tips in the creation of a proper database that makes it amenable for subsequent analysis.

3.7 Common Mistakes to Avoid

3.7.1 Questionnaire Design

1. Avoid creating answers in questionnaires that do not match the questions!

   For example, question on frequency of smoking with responses such *as* 'never', 'last week' and 'last month', which are measuring recency instead of frequency. Pilot testing the questionnaire may help to identify such issues.

2. Do not use questions that are double negative or double barrelled.

   For example, 'Have you been unsuccessful in the past 12 months in terms of quitting smoking?' This is a double-negative question. Another example is: 'Are you currently not pregnant, not lactating and practicing adequate contraception?' Options: 'Yes'/'No'. This is a triple-barrelled and double-negative question.

3. Not providing the unit of measurement in the questionnaire

   For example, 5 kg or 5 £ can make a huge difference in the analysis of neonatal weight data. An equally bad scenario is when the units are not standardized when measured across wards or by different data entry personnel (e.g. number of falls without considering the number of patients in each ward).

4. Not providing enough boxes for numerical responses as well as not indicating where the decimal point should be in a questionnaire.

   This creates inconsistencies in data entry and may result in severe bias in the results of the analysis and make interpretation of the data difficult.

5. Overusing the 'Others, pls specify' option

   This happens especially when the data entry person is unsure of the correct option to select among the pre-codes, usually when reporting adverse events. Researchers tend to pick 'Others, pls specify' when there are other appropriate options, thus making it difficult to assess the safety profile of the drug without recoding the data. For example, when patient complains of frequent exhaustion, researcher codes it as 'Others, pls specify' instead of the given code 'Fatigue'.

6. Keying in dates in various mixed formats

   For example, 22 Jan 2012, 31/01/13, 3-06-12, and so on. It is not clear which number is the day and which is the month in the latter example. It will be very difficult to rectify this mistake once the database has been completed. It is advisable to use Excel's column formatting tool to ensure date formats are fixed before data entry is undertaken.

3.7.2 Database Design

1. Using colour codes to differentiate data cell values

   For example, 'blue' for males and 'red' for females. As mentioned previously, statistical software packages (mostly) are colour blind and may not be able to differentiate the data.

2. Inconsistencies in representing missing data

   For example, keying in '99', 'NA', '888', and so on in the same database to represent missing values. Statistical software packages have different ways to identify missing data. Incorrect use of numbers to identify missing data with codes may bias the analysis severely. For example, keying in 888 for missing data as identifying code for data such as length of stay. It is best to leave the cell empty instead if unsure.

3. Inconsistencies in keying in text fields

   For example, keying in text data in both upper case and lower case, that is 'Male', 'male' and 'MALE'. Most statistical software will treat these as distinct entries, and this will create unnecessary work later on to recode them into a single category. It is advisable to use pre-code numbers (e.g. '1' for males and '0' for females).

4. Keying in units of measurement with numeric data (e.g. <5 mm, 6 mm*, etc.)

   Some clinicians like to key in the units of measurements in the same column as the data, resulting in an alphanumeric field. Most software will treat the entire column as a string variable (even if one of the cells has an alphabet in it), and it may not be possible to calculate descriptive statistics such as the mean and median as the column is treated as a string variable. Units of measurement should be stored separate to the data, perhaps in a data dictionary or in the labelling of the variable.

5. Including unnecessary variables in the dataset

Clinicians sometimes key in patient names and other identifiers that are not used in the analysis. This should be avoided, especially when sharing the data with others (e.g. biostatisticians). If they are necessary (e.g. for use as unique identifiers in merging with other datasets), a master list of names and code numbers should be created and stored separately in a secure location.

6. Creating additional information in the Excel datasheet

Summary statistics such as the mean and standard deviation are sometimes calculated using Excel's in-built statistical functions and displayed at the end of columns and rows to provide an overall summary of the data. Sometimes, additional fields not used in the analysis are also included in columns/rows, and these fields are hidden. Data may also be partially sorted on specific columns. All these actions should be avoided as they may pose problems when importing the data into a statistical software.

3.8 Tools and Resources

1. A step-by-step guide on how to create a database, enter data and format it in Excel. http://spreadsheets.about.com/od/datamanagementinexcel/ss/excel_database.htm. Date accessed: 1 August 2013.

2. This website lists simple steps one can take to minimise data entry errors in using Excel. http://www.cogniview.com/blog/avoiding-costly-and-dangerous-data-entry-disasters/. Date accessed: 1 August 2013.

3. An excellent resource from the University of Nebraska on preparing data in Excel. http://www.unmc.edu/publichealth/ccorda_hlp_excel. htm. Date accessed: 1 August 2013.

4. The Patient Reported Outcomes Measurement Information System (PROMIS) is funded by the National Institutes of Health (NIH), and the website contains a system of highly reliable, valid, flexible, precise, and responsive assessment tools that measure patient-reported health status: http://www.nihpromis.org/. A summary of publications of the PROMIS instruments is provided at http://www.nihpromis.org/science/publications. Date accessed: 3 August 2015.

3.9 A Collaborative Case Study

Project title. Dissociative semantic breakdown in Alzheimer's disease (AD): Evidence from multiple category fluency test (Ting et al., 2013).

Collaborating clinician. Dr. Simon Ting Kang Seng, MBBS, MRCP (London); consultant neurologist, National Neuroscience Institute, Singapore.

Study summary. This study analysed the baseline neuropsychological assessment performance of food and animal fluency test of 296 AD patients who presented to Singapore General Hospital, Singapore, from 1995 to 2006, and agreement assessed using the kappa statistic. Fair-to-moderate agreement between food and animal category fluency test was found especially in the mild AD cases (Literate: kappa 0.40; Illiterate: kappa 0.42). The study also found that agreement level significantly increased when the disease progressed.

Study dataset. The original database was presented in an SPSS format with variables well coded and labelled. This dataset was subsequently imported into Stata format using Stat/Transfer software and analysed. Figure 3.7 shows how the database was well organised with categorical variables coded. Initial preliminary analysis showed that for

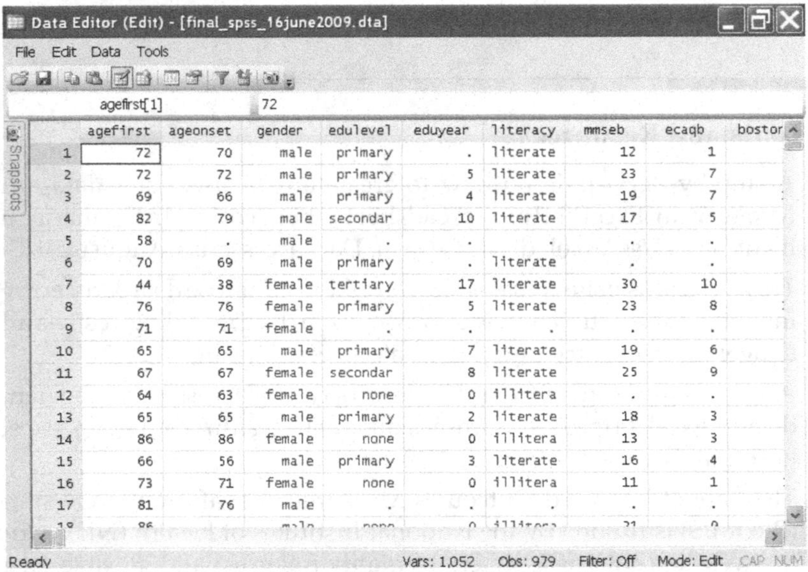

	agefirst	ageonset	gender	edulevel	eduyear	literacy	mmseb	ecaqb	bostor
1	72	70	male	primary	.	literate	12	1	
2	72	72	male	primary	5	literate	23	7	
3	69	66	male	primary	4	literate	19	7	
4	82	79	male	secondar	10	literate	17	5	
5	58	.	male	.	.	.	.	.	
6	70	69	male	primary	.	literate	.	.	
7	44	38	female	tertiary	17	literate	30	10	
8	76	76	female	primary	5	literate	23	8	
9	71	71	female	.	.	.	.	.	
10	65	65	male	primary	7	literate	19	6	
11	67	67	female	secondar	8	literate	25	9	
12	64	63	female	none	0	illitera	.	.	
13	65	65	male	primary	2	literate	18	3	
14	86	86	female	none	0	illitera	13	3	
15	66	56	male	primary	3	literate	16	4	
16	73	71	female	none	0	illitera	11	1	
17	81	76	male	.	.	.	.	.	

FIGURE 3.7
Sample Stata database.

two of the key variables – food and animal fluency scores – there were two extremely large outlier values of 1999 and 88. After consulting with the principal investigator and him checking the original forms again, it was determined that the number 1999 was an error and thus coded as missing value, and that 88 was a typing error that was meant to be 8. The exercise demonstrated that even with a well-collected and well-organised dataset, it is possible to have data entry errors, and it would be prudent to conduct a preliminary check of the data prior to the analysis.

The other point to make is that the database should include variables in their original form. Derived variables that are needed in the study should be created in the analysis stage by writing some simple software codes. For example in this study, the clinician needed new variables for food and animal fluency scores grouped into normal and abnormal values, as well as stratified by educational levels. This was achieved by writing and executing the following commands in Stata:

```
ge anib_normal_primary=(anib>=7) if edu_gp==1 & anib~=.
ge anib_normal_secondary=(anib>=9) if edu_gp==2 & anib~=.
ge foodb_normal_primary=(foodb>=8) if edu_gp==1 & foodb~=.
ge foodb_normal_secondary=(foodb>=11) if edu_gp==2 & foodb~=.
```

In the above commands, `anib` and `foodb` are the initial animal and food continuous scores and `edu _ gp` is a categorical variable for educational level (1: 'primary level and below' and 2: 'secondary level and above').

Key Learning Points

1. Learn good practices in questionnaire design.
2. Understand and be able to explain the different types of variables.
3. Distinguish between various scales of measurements.
4. Describe how to set up a simple but effective database for research using Excel.
5. Improve the efficiency process of discussing with your collaborators on the design of the data collection form and database by utilising the list of 'key questions to ask' checklist.
6. Avoid the pitfalls in questionnaire and database design.

References

Bellary, S., Krishnankutty, B., & Latha, M. S. (2014). Basics of case report form designing in clinical research. *Perspect Clin Res, 5*(4), 159–166. doi:10.4103/2229-3485.140555.

Bland, J. M., & Altman, D. G. (1986). Statistical methods for assessing agreement between two methods of clinical measurement. *Lancet, 1*(8476), 307–310.

Clark, G. T., & Mulligan, R. (2011). Fifteen common mistakes encountered in clinical research. *J Prosthodont Res, 55*(1), 1–6. doi:10.1016/j.jpor.2010.09.002.

Hulley, S. B., Cummings, S. R., Browner, W. S., Grady, D. G., & Newman, T. B. (2007). *Designing Clinical Research*. Philadelphia, PA: Lippincott Williams & Wilkins.

Sperber, A. D. (2004). Translation and validation of study instruments for cross-cultural research. *Gastroenterology, 126*(1 Suppl. 1), S124–S128.

Streiner, D. L., & Norman, G. R. (1995). *Health Measurement Scales: A Practical Guide to Their Development and Use* (2nd ed.). New York: Oxford University Press.

Ting, S. K., Hameed, S., Earnest, A., & Tan, E. K. (2013). Dissociative semantic breakdown in Alzheimer's disease: Evidence from multiple category fluency test. *Clin Neurol Neurosurg, 115*(7), 1049–1051. doi:10.1016/j.clineuro.2012.10.030 S0303-8467(12)00564-1 [pii].

Weiner, E. A., & Steward, B. J. (1984). *Assessing Individuals*. Boston, MA: Little Brown.

4

Sample Size and Power Calculations

4.1 Introduction

Sample size calculation is one of the most commonly performed tasks of a collaborating biostatistician. Sample size and power calculations are essential in any form of research that involves hypothesis testing. This encompasses studies of the various forms of study designs discussed in Chapters 1 and 2. If too few patients are recruited, there is a possibility of the study being under-powered to detect any meaningful effect, and conversely if too many patients are recruited, the study can be over-powered, and this can then result in wastage of time and money, potentially putting patients in the study at additional risk, as well as causing unnecessary delays in the study (Earnest, 2010). Declaring a trivial effect size statistically significant can lead to subsequent confusion with clinical significance. Increasingly, journal submissions request for sample size calculations to be explicitly stated in the methods section of the manuscripts, and articles are often sent for external statistical review. Some journals also have statisticians as members of their editorial boards, and this highlights the emphasis that is placed on statistical methodological issues, including appropriate statistical analysis plan and sample size calculations.

International guidelines such as the Consolidated Standards of Reporting Trials (CONSORT) (Moher et al., 2012) statement for randomised controlled trials (RCTs) and strengthening the reporting of observational studies in epidemiology (STROBE) (von Elm et al., 2007) checklist for observational studies such as cohort and case–control study designs place importance in the issue of sample size by including it as one of the criteria, and specifically asking for detailed information on how sample size was calculated to be provided. The guidelines also provide carefully worded examples on sample size calculations.

Sample size/power calculations are equally important and requested for by a grant review committee, which sometimes includes a biostatistician as one of the panel members. In fact, this author sat on a local review panel of the National Medical Research Council's Clinician Scientist-Individual Research Grant (CS-IRG) for 2 years and experienced how often the panel turned to the statistician to comment on the sample size calculations for many of the individual grant proposals that were being considered for funding!

This chapter will highlight the key ingredients in a sample size calculation as well as provide the crucial link with hypothesis testing. Clinical examples will be used to demonstrate some of the commonly performed sample size calculations in clinical research. Rather than showcasing formulas and the theoretical derivation of the sample size equations, this chapter will exemplify sample size calculations using Sample Size Tables for Clinical Studies Software (Machin et al., 2009), PS Power and Sample Size (PSPSS) calculations (Dupont & Plummer, 1990) (http:// biostat.mc.vanderbilt.edu/wiki/Main/PowerSampleSize. Date accessed: 12 August 2013) as well as Stata software. Section 4.5 includes some key questions to ask collaborators when deliberating sample size calculations, and hopefully such a checklist would come in useful for a biostatistician who is meeting the collaborator and undertaking sample size calculations for the first time. Finally, Section 4.6 discusses some common mistakes to avoid in sample size calculations, which help biostatisticians identify the common pitfalls when undertaking such a task.

4.2 Linking Hypothesis Testing and Sample Size

Sample size calculation should be performed for each hypothesis listed under the primary objective(s) of the study. Each hypothesis is then linked to the aim as described in Section 1.4. The following section describes the relationship between sample size/power and hypothesis testing (Earnest, 2010). The power of a statistical test is defined as the probability of rejecting the null hypothesis when the alternative hypothesis is actually true. Consequently, we want power to be large, and traditionally a value of 0.80 is often used in most studies. Power is closely related to type 2 error, with power = 1 − prob(type 2 error). Hence, when one defines power to be 0.90 for a particular statistical test, there is a .10 chance of a type 2 error (β) occurring, that is, the probability of rejecting the alternative hypothesis when it is true is .1. Power also increases when the type 1 error (or level of significance) increases. The level of significance is the proportion of time the test will reject the null hypothesis when it is true. Level of significance is denoted as α and usually set at 0.05. Eighty percent power with 0.05 level of significance would mean a type 2 error (0.20), which is 4 times the type 1 error (0.05). Figure 4.1 shows the relationship between hypothesis testing and the two types of errors.

	H_0 true	H_0 false
Reject H_0	Type I error (α)	True positive
Do not reject H_0	True negative	Type II error (β)

FIGURE 4.1
Relationship between hypothesis testing and type 1 and 2 errors.

4.3 Ingredients in a Sample Size Calculation

It is important to understand the mechanism behind sample size calculations so that the right questions can be directed to a clinician/researcher to ensure that sample size calculations are done correctly and efficiently. Knowing the components of sample size calculation is also important in the design of a study as it allows one to strategize and plan the study in the most efficient way. For instance, one might try and keep the population more homogeneous (e.g. by restricting the age groups) in order to reduce the variability in the data, and hence requiring fewer subjects for the experiment. The following Sections 4.3.1 through 4.3.7 are key elements of a typical sample size calculation for a clinical research project.

4.3.1 Objective of the Study

If the aim of the study was to estimate a parameter instead of testing a hypothesis, the considerations for sample size determination would be different. For example, if the aim of the study was to estimate the prevalence of chronic obstructive pulmonary disease in a community-based study, a key consideration in the sample size determination would be the level of precision required to estimate the prevalence. A larger sample size would provide for narrower 95% confidence intervals. As shown in Figure 4.2, we can observe that increasing the sample size tenfold from 90 to 900 results in a narrower 95% confidence interval of (0.52–0.59). However, if there is a specific hypothesis being tested, then other considerations such as type 1 and type 2 errors, the nature of the primary outcome measure, the variability in the data and the minimum effect size that needs to be detected will be other important factors.

`cii 90 50`

Variable	Obs	Mean	Std. Err.	— Binomial Exact — [95% Conf. Interval]	
	90	.5555556	.0523783	.446996	.6603558

`cii 900 500`

Variable	Obs	Mean	Std. Err.	— Binomial Exact — [95% Conf. Interval]	
	900	.5555556	.0165635	.5223989	.5883474

FIGURE 4.2
Relationship between sample size and precision.

4.3.2 Type 1 Error (Level of Significance)

As mentioned previously, for sample size calculations, type 1 error is usually set at 0.05. The larger this threshold, the smaller the sample size required in a study (Figure 4.3). As we can see, the sample size increases from 37 to 47 when the level of significance changes from 0.10 to 0.05. In some RCTs involving new drugs, especially those with potentially harmful side effects, researchers may want to determine that the drug is really efficacious before terminating the study, and hence they need to provide justification to set a more stringent level of significance of (say) 0.01.

4.3.3 Type 2 Error (1 – Power)

In sample size calculations, power is usually set at 0.80 and sometimes 0.90. This gives rise to type 2 error rates of 0.20 and 0.10, respectively. All other things being equal, there is usually a trade-off between type 1 and type 2 errors in sample size calculations. Another point to note is that sometimes power calculations are done *post hoc* or retrospectively after a study is completed. This is usually at the request of journal reviewers who feel that the study may have been under-powered to detect any meaningful effect. Sometimes, power calculations are performed for large routinely collected data sources to see whether it is feasible to perform any subgroup or stratified analysis.

4.3.4 Variability

The larger the variability, the greater the sample size required in the study. This direct relationship is shown by the following formula (Pocock, 2000) for an example of comparing means in two groups, where *n* is the number of

. `sampsi 0.5 0.7, power(0.8) onesample alpha(0.1)`

Estimated sample size for one-sample comparison of proportion
to hypothesized value
Test Ho: p = 0.5000, where p is the proportion in the population

Assumptions:

        alpha =   0.1000 (two-sided)
        power =   0.8000
  alternative p =   0.7000

Estimated required sample size:

                n =   37

. `sampsi 0.5 0.7, power(0.8) onesample alpha(0.05)`

Estimated sample size for one-sample comparison of proportion
to hypothesized value
Test Ho: p = 0.5000, where p is the proportion in the population

Assumptions:

        alpha =   0.0500 (two-sided)
        power =   0.8000
  alternative p =   0.7000

Estimated required sample size:

                n =   47

FIGURE 4.3
Relationship between sample size and type 1 error.

patients in each group; $f(\alpha,\beta)$ is a function of type 1 and 2 errors, respectively; σ^2 is the variance and μ_2 and μ_1 are the means in the two groups:

$$n = \frac{2\sigma^2}{(\mu_2 - \mu_1)^2} \times f(\alpha,\beta) \qquad (4.1)$$

There are several strategies one can adopt in the design of a study to reduce variability in the data, such as using more precise instruments, taking multiple observations from the same patient and also selecting a more specific target population. However, selecting specific populations will have an effect on the generalisability of the study results and often a trade-off is required. Often, inclusion and exclusion criteria are used in studies to exclude potential confounders in the data and to ensure that the study sample is homogeneous. One needs to be careful not to exclude the population under study by inadvertently excluding them in the exclusion criteria (e.g. excluding those above 40 years who are more predisposed to be hypertensive in a blood pressure reduction trial). The variance estimate is usually obtained from previously published studies, although these approximations have been suggested as underestimates as they lead to higher probabilities of finding statistically significant results, and hence a higher chance of the paper is published (Wittes, 2002). Pilot studies are alternative sources of variance estimates.

4.3.5 The Effect Size

In a clinical research project, the effect size relates to how big or small a difference we observe in the outcome variable. Effect size measurements vary according to the objectives of study and outcome measures and include standardised difference in means and proportions, odds ratios and hazard ratios. For instance, a standardised mean would involve dividing the difference in the mean outcome between groups by the corresponding standard deviation. A clinically meaningful effect size relates to an effect that, when observed at the end of the study, would be deemed useful enough to warrant a change in clinical practice or introduction of an intervention that is being studied in the experiment. This effect size needs to be specified before an experiment is conducted and is often the most difficult information to obtain from the clinician during a discussion on sample size calculation. Often, this information is gleaned from related publications in the field looking at the specific intervention and outcome. In the absence of such publications, a pilot study is recommended to get such information. Sometimes, researchers use standardised estimates of effect sizes in sample size calculations such as Cohen's classification of small (0.2), medium (0.5) and large (0.8) effect sizes (Cohen, 1988, 1992). Generally, to estimate a small effect size, a large sample size is required.

4.3.6 Two-Sided versus One-Sided Tests

Most experiments involve hypothesis testing which is two sided (i.e. the alternate hypothesis can go in either direction from the null hypothesis). However, one-sided tests state *a priori* that the alternate hypothesis can only take one direction. This is often difficult to ascertain before the experiment is conducted and should be avoided. One-sided tests require smaller sample size compared to two-sided tests. As we can see from Figure 4.4, when we calculate the sample size for a two-sided test (figure on left), the number is 108 per arm. However, this number drops substantially to 83 for a one-sided test with all other parameters kept the same. Although it is tempting to specify one-sided tests because it provides a smaller sample size, strong justification should be elicited from the collaborator and documented before this is done.

4.3.7 Other Considerations

In any sample size calculation, it would be prudent to account for some level of dropouts or non-completion in the study. This is especially important in a longitudinal study or when the anticipated dropout rate is going to be high, and thus may cause the study to be under-powered if the dropouts are not accounted for. If the anticipated dropout rate, r, can be ascertained, then the final sample size can be multiplied by a simple correction factor $1/(1 - r)$ to account for potential dropout or non-completers.

Non-compliance is another potential problem, particularly in RCTs. For instance, a patient assigned to a new drug may stop taking the drug after a while, or a patient on placebo may need to be started on therapy midway through the trial, and this will have an obvious impact on the event rates in the two groups. One possible solution is to inflate the sample size initially for this 'non-compliance' using the factor $(1 - p_t)^2$, where p_t is the proportion of subjects in the treatment group who have stopped taking the new drug (Wittes, 2002).

If there are subgroups that are going to be analysed or if the analysis needs to be stratified by key covariates, these would also increase the overall sample size in the study. Also, complex sample size calculations exist for more sophisticated study designs such as cluster randomised trials, adaptive phase 1 clinical trials and equivalence trials, and appropriate resources should be consulted to review the sample size calculations in these situations.

4.4 Commonly Performed Sample Size Calculations

In this section, we highlight some commonly performed sample size calculations using different software such as PSPSS, which is free software available at http://biostat.mc.vanderbilt.edu/wiki/Main/PowerSampleSize (Date

. sampsi 5.5 6.1, sd1(1.5) a(0.1)

Estimated sample size for two-sample comparison of means

Test Ho: m1 = m2, where m1 is the mean in population 1
                  and m2 is the mean in population 2

Assumptions:

```
    alpha =   0.1000  (two-sided)
    power =   0.9000
       m1 =   5.5
       m2 =   6.1
      sd1 =   1.5
      sd2 =   1.5
    n2/n1 =   1.00
```

Estimated required sample sizes:

```
       n1 =     108
       n2 =     108
```

. sampsi 5.5 6.1, sd1(1.5) a(0.1) onesided

Estimated sample size for two-sample comparison of means

Test Ho: m1 = m2, where m1 is the mean in population 1
                  and m2 is the mean in population 2

Assumptions:

```
    alpha =   0.1000  (one-sided)
    power =   0.9000
       m1 =   5.5
       m2 =   6.1
      sd1 =   1.5
      sd2 =   1.5
    n2/n1 =   1.00
```

Estimated required sample sizes:

```
       n1 =      83
       n2 =      83
```

FIGURE 4.4
Sample size implications for two-sided versus one-sided tests.

accessed: 27 November 2014), and in-built sample size programs within Stata. Clinical problems will be used to exemplify the calculations. PSPSS is menu driven and interactive, so it is relatively easy to use. One can also click on the input parameter that is hyperlinked to the help file for more information. For example, clicking on α will bring up a window with the following help text 'The Type I error probability for a two-sided test. This is the probability that we will falsely reject the null hypothesis'. The software also provides a log file for each calculation, which describes details of the sample size calculations, such as the level of significance, power and effect size specified in the calculation. This text can be copied into manuscripts or proposals, although the generic wordings need to be changed according to the specific study details.

4.4.1 Comparing Proportions between Two Independent Groups

In study designs in which the key outcome measure is dichotomous (e.g. dead/alive) and exposure variable consists of two groups (e.g. drug X vs. placebo), one may wish to calculate the sample size in each group (n), given probability of death in each groups, p_0 and p_1, two-sided test size α, power and the ratio of control to experimental subjects, denoted as m. Suppose in an RCT, $p_0 = 0.2$ and $p_1 = 0.1$, two-sided test size $\alpha = 0.05$, power $= 0.8$ and $m = 1$. After launching PSPSS, one needs to click on the 'Dichotomous' tab, key in all relevant information and parameters as above, and click on 'Calculate'. The sample size per group, n, is then calculated to be 199 (Figure 4.5). It may be possible to recruit more patients in the intervention group compared to the placebo so as to encourage more subjects to participate in the study, but this will result in a much larger study (e.g. 3:1 allocation will increase sample size by 33%) (Wittes, 2002).

4.4.2 Comparing Means between Two Independent Groups

In a study in which the outcome is continuous in nature (e.g. length of stay [LoS]) and the exposure variable is dichotomous (e.g. presence of comorbid such as ischemic heart disease [IHD] measured as yes/no variable), the formula for the sample size in each group is given by Equation 5.2 (Machin et al., 2009), where two-sided test size is α, allocation ratio between the two groups is φ (denoted m in software) and the difference in means between two groups is δ, along with the common standard deviation of σ. Suppose in a hospital-based cross-sectional study, $\alpha = 0.05$, $m = 1$, power is 0.8, the difference in length of stay between those with and without IHD, $\delta = 1$ day and the standard deviation for LOS, $\sigma = 2$, one needs to select the 't-test' tab in PSPSS, then the sample size per group would be calculated to be 64 (Figure 4.6).

4.4.3 Estimating Hazard Ratios in a Survival (Time to Event) Analysis

In a survival analysis, the main outcome is the survival time till a particular event (usually death). There are several considerations for sample size

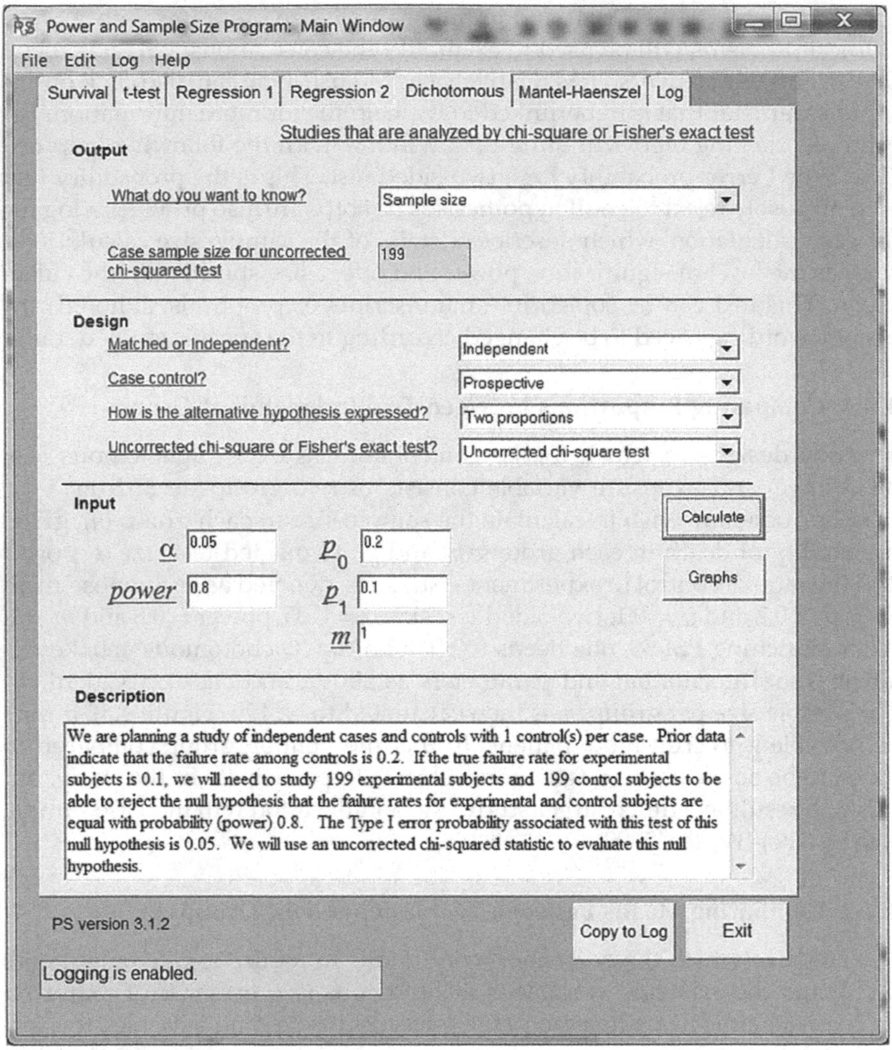

FIGURE 4.5
Sample size calculations for comparing proportions between two independent groups.

determination as described in the PSPSS software (Dupont & Plummer, 1990): for a given level of significance α and power $1 - \beta$, m_1 is the median survival time on control treatment, m_2 is for the experimental treatment group, R is the hazard ratio, A is the accrual time during which patients are recruited, F is the additional follow-up time after end of recruitment and m is the ratio of control to experimental patients. Sample size calculation for survival analysis is found in the 'Survival' tab in PSPSS. Suppose in an RCT, a neurologist wants to evaluate the effectiveness of a new drug in treating

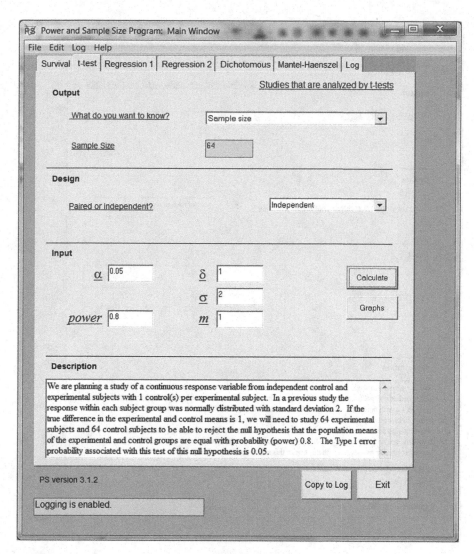

FIGURE 4.6
Sample size calculations for comparing means between two independent groups.

patients with acute stroke compared to standard drugs. Suppose that the current survival among those taking standard drugs is 4 years, and that the investigator wishes to recruit patients over a 3-year study period with a further follow-up of 5 years from last recruitment. Assuming that pilot study indicated treatment effect of a hazard ratio of 1.5, the number of subjects that need to be recruited per arm would be 162 (Figure 4.7). The total sample size would be 324. After accounting for an estimated 20% dropouts, 360 patients would need to be recruited for the trial.

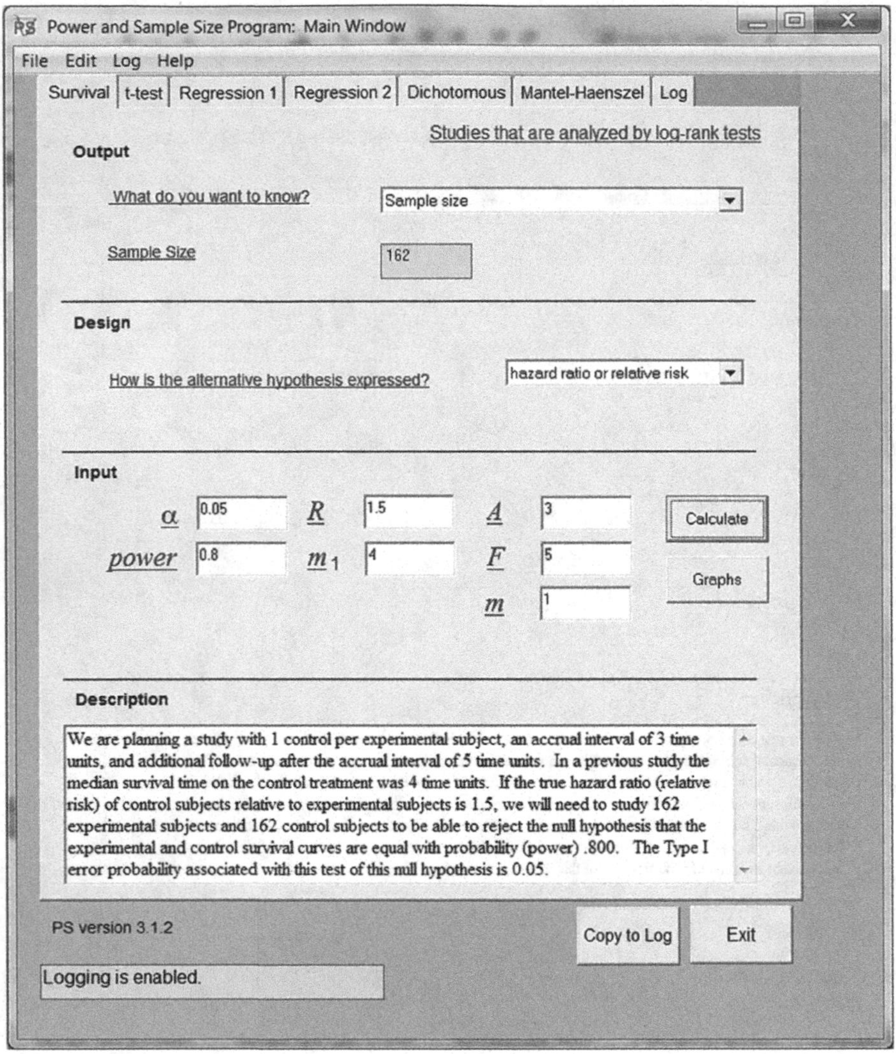

FIGURE 4.7
Sample size calculations for estimating hazard ratios.

4.4.4 Estimating Coefficients in a Linear Regression Model

The ordinary least squares regression model is one of the most common statistical techniques performed in clinical research. It is used when the outcome is measured on a continuous scale (e.g. length of stay, systolic blood pressure, etc.), but the predictor variables can be either categorical or continuous in nature. Sample size calculations for linear regression models take the following into consideration (Dupont & Plummer, 1990): level of significance α, power $1 - \beta$, standard deviation of the regression errors σ, standard deviation of the predictor variable σ_x and the regression slope or effect one wishes to detect λ. In a clinical research project, suppose a geriatrician wanted to estimate the relationship between the age and the length of stay among patients attending his/her outpatient clinic over the past 6 months. Based on a previous small pilot data, he/she estimated that $\sigma = 5$, $\sigma_x = 0.5$ and $\lambda = 2$. The minimum number of patients he/she needs to recruit to demonstrate a statistically significant result would be 198 (Figure 4.8). The calculation done is the 'Regression 1' tab of PSPSS. The software also allows one to change treatment levels and estimate σ from the standard deviation of the outcome

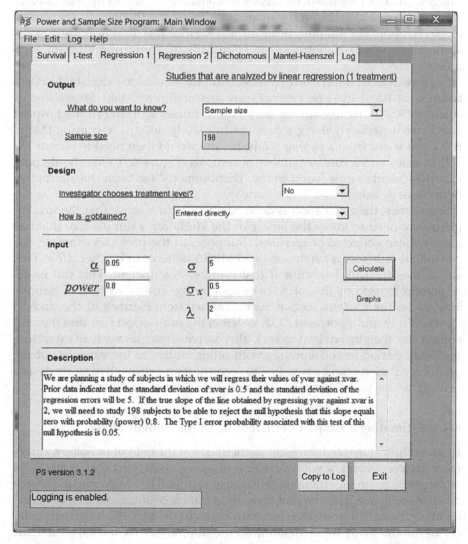

FIGURE 4.8
Sample size calculations for a linear regression model.

variable along with the correlation coefficient between the outcome and predictor variables, but these parameters are best obtained from a pilot study.

4.4.5 Estimating Odds Ratios in a Logistic Regression Model

The logistic regression model is the next most popular statistical technique used in clinical research. It is particularly used to analyse outcomes that are dichotomous in nature (e.g. dead vs. alive, hypertension—yes/no, etc.). Once again, the predictor or independent variables can be continuous or categorical. The biostatistician needs to consider the following parameters in a sample size calculation for an analysis involving the logistic regression model (Dupont & Plummer, 1990): level of significance α, power $1 - \beta$, probability of exposure in controls p_0 and clinically meaningful minimum odds ratio to detect ψ.

Suppose that an infectious disease consultant wanted to evaluate the effectiveness of hand hygiene practices on methicillin-resistant *Staphylococcus aureus* (MRSA) carriage among doctors and nurses working in the hospital wards retrospectively using a case–control study design. Assuming that p_0 is 0.2 and ψ is 2 from a similar study, he/she would then need to recruit 172 MRSA cases and a similar number of controls (Figure 4.9). One should note that this calculation is found in the 'Dichotomous' tab with the alternative hypothesis selected as an 'Odds Ratio'.

Sometimes, the calculation is done backwards in a sense that the investigator may need to know the power of the study for a sample size that has already been collected or recruited. Suppose, in the previous example, the consultant only sees a maximum of 50 MRSA patients in his/her clinic. He/she wishes then to determine if that number is sufficient. This can easily be accommodated by the software by specifying 'Power' instead of 'Sample Size' under the 'Output' section. As we can see from Figure 4.10, the study is then severely underpowered (0.32) to detect the same effect size, and the consultant can then be advised to seek alternative strategies such as extending the study period or collaborating with other centres to increase the subject pool. Calculations can be tweaked to account for a prospective study design instead, where the relative risk is specified instead of an odds ratio.

4.4.6 Estimating a Kappa Coefficient in an Agreement Study

In some clinical research projects, particularly in the field of radiology, agreement between raters or within raters is often assessed, particularly for X-ray or magnetic resonance imaging (MRI) scan results. The kappa statistic is usually used to quantify the level of agreement after accounting for chance agreements. In Stata, a user-written program exists (type 'findit sskdlg' under the command field), which determines the sample size for the kappa statistic measure of inter-rater agreement when there are two unique raters evaluating a binary event. Although the following Section 4.4.7 (and Chapter 5) make

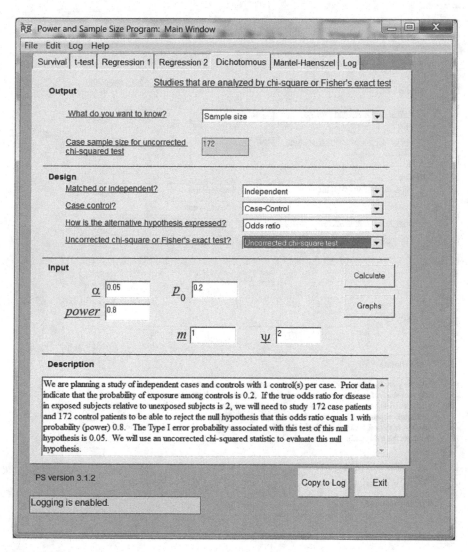

FIGURE 4.9
Sample size calculations for a logistic regression model.

use of Stata to demonstrate sample size calculations and the interpretation of statistical output, it is not the purpose of this book to be an instructional manual for the use of Stata. There are excellent resources out there which are more than capable of demonstrating this, but for simple instructions on the input and import of data into Stata, the following notes and movie from the University of California, Los Angeles (UCLA) may be helpful (http://www. ats.ucla.edu/stat/stata/notes/default.htm. Date accessed: 25 November 2015).

When the command 'sskdlg' is typed, an interactive menu will be displayed. The sample size is chosen based on precision estimate considerations,

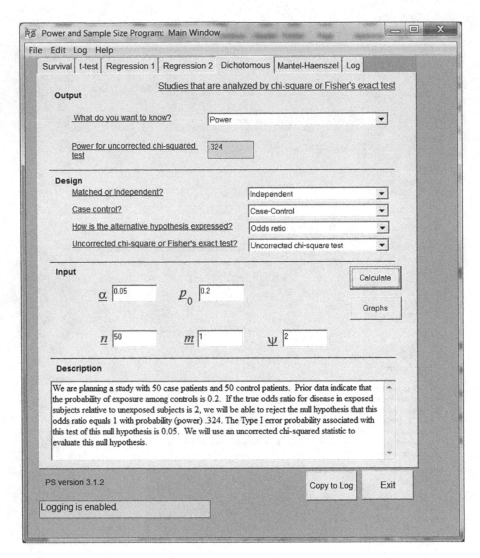

FIGURE 4.10
Power calculations for a logistic regression model.

such that the standard error of the estimate and the ensuing limits for a $100(1 - \alpha)\%$ confidence interval do not exceed stipulated values, with variance calculation details shown here (Fleiss et al., 1969). Users need to input the following five parameters for the sample size calculation: expected kappa κ; proportion of positive results rated by the first- and second observers p_1 and p_2; absolute precision d, which is the minimum envisaged difference between κ and its lower (or upper) confidence limit; and finally the confidence levels (usually set at 95%).

In this example of a hypothetical clinical services audit, the radiology department wanted to assess the agreement between two radiologists, in terms of classifying MRI results into positive and negative results for brain tumour. Suppose that pilot data indicated that the expected κ was 0.8, and p_1 and p_2 were expected to be 0.2 and the precision was set at 5%, as shown in Figure 4.11, the total number of scans that would need to be read by both radiologists would be 885.

4.4.7 Repeated Measures Analysis

In hospital-based records or registries, it is common to see measurements for the same patient taken several times. Examples include quality of life (QoL) measured during each visit to the outpatient clinic, or serial daily measurements of blood pressure or laboratory tests performed during admission to hospital and so on. There are established statistical methods available to analyse such data (e.g. generalised estimating equations, random effect models, etc.). For sample size calculation, efficiency is achieved in accounting for the within-patient correlation in the data. Stata has an in-built command for calculating the sample size for repeated measures analysis.

The command is as follows: sampsi m1 m2, sd1(#) sd2(#) method(change) pre(#) post(#) r1(#), where m1 and m2 are the pre- and post-intervention mean values, sd1 and sd2 indicate the standard

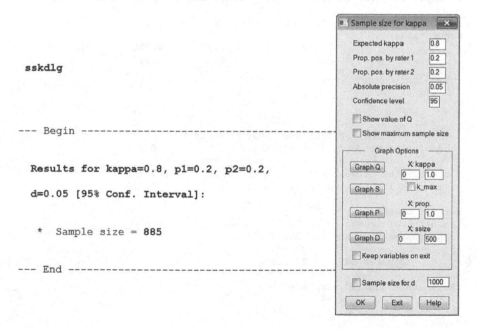

FIGURE 4.11
Sample size calculations for estimating kappa.

deviation estimates for the pre- and post-intervention, pre and post are the number of pre- and post-measurements done and r1 indicates the correlation between follow-up measurements, respectively.

An internal medicine specialist wanted to evaluate a particular care path way, an intervention hypothesised to reduce systolic blood pressure levels among his/her patients. He/she would like to take three systolic blood pressure measurements before and after the implementation of the care path way. Based on the previous data, he/she estimates that the intervention would reduce the mean systolic values from 145 to 135 mmHg with standard deviation estimates of 50 and 45 in the two periods, and a within-person correlation of 0.7. Based on sample size calculations in Stata (Figure 4.12), 96 patients would need to be recruited in the study.

4.4.8 Cluster Randomised Trials

There is an additional complexity in designing cluster randomised trials, which was introduced in Section 2.5.1. Patients within the same cluster would exhibit similarity in terms of their response to the outcome measured, and this is captured as a within-cluster dependence, and the relative contribution to overall variance is quantified by the intra-cluster correlation coefficient (ICC) or ρ. Based on the ICC, the design effect (DE) is calculated as shown

```
. sampsi 145 135, sd1(50) sd2(45) method(change) pre(3) post(3) r1(0.7)

Estimated sample size for two samples with repeated measures
Assumptions:
                                              alpha =   0.0500  (two-sided)
                                              power =   0.9000
                                                 m1 =      145
                                                 m2 =      135
                                                sd1 =       50
                                                sd2 =       45
                                              n2/n1 =     1.00
                number of follow-up measurements =        3
   correlation between follow-up measurements =    0.700
              number of baseline measurements =        3
   correlation between baseline measurements =    0.700
     correlation between baseline & follow-up =    0.700

Method: CHANGE
   relative efficiency =      5.000
      adjustment to sd =      0.447
          adjusted sd1 =     22.361
          adjusted sd2 =     20.125

 Estimated required sample sizes:
                 n1 =          96
                 n2 =          96
```

FIGURE 4.12
Sample size calculations for repeated measures analysis.

in Equation 4.2 (Batistatou et al., 2014), where m is the number of patients in each cluster and $\bar{m}$ is the anticipated mean cluster size. The sample size is firstly calculated using the usual methods described in Sections 4.4.1 through 4.4.7. Subsequently, the sample size is inflated by multiplying it with the DE to account for clustering.

Suppose in the trial example discussed in Figure 4.5, instead of an individual RCT, the investigators decided to perform a cluster RCT at the local government area in Melbourne suburbs instead. Assuming that ρ is 0.05, the mean sample size per cluster is 20 with SD = 3, then DE = 1 + (20 + (9/20)−1) × 0.05 = 1.9725. The total sample size would then be 1.9725 × 199 × 2 = 786.

$$DE = 1 + \left(\bar{m} + \frac{SD(m)^2}{\bar{m}} - 1 \right) \rho \qquad (4.2)$$

Obviously, the ICC would have a profound impact on the final sample size and needs to be selected carefully. The University of Aberdeen, Aberdeen, has published a spreadsheet of ICCs, which were collected from published studies with a range of study designs such as RCTs, cross-sectional studies and hospital audit data. The collection covers a range of diseases such as coronary heart disease, asthma, hypertension and epilepsy can be used for prospective sample size calculation. The website can be accessed at http://www.abdn.ac.uk/hsru/research/delivery/behaviour/methodological-research/ (Date accessed: 28 November 2014).

4.5 Key Questions to Ask a Clinician

1. What is the primary aim and associated hypothesis in your study?

   Normally, sample size calculation is performed for the primary aims of the study, and if the collaborator has too many listed, he/she can be advised to shift some of the not-so-important outcomes to the secondary or even exploratory aims.

2. What is the nature of your outcome and explanatory variables?

   The nature of variables will determine the specific hypothesis that is to be tested, and hence the type of sample size calculation needed. For instance, if one wishes to look at the relationship in the length of stay (continuous outcome variable) between stroke and dementia patients (binary explanatory variable), then an independent Student's t-test may be appropriate, and the relevant sample size calculation can be performed.

3. Is the comparison between or within subjects?

   Within-subject comparisons usually result in smaller sample size calculations. However, repeated measures will require more sophisticated sample size computations, as these require more information such as estimates of within-person correlations.

4. What would be the minimum clinically meaningful effect size you wish to detect?

   This is usually the most difficult information to get from the collaborator, and it is not a good idea to run a series of sample size calculations based on several plausible values of the effect size, and let the collaborator choose the most convenient one. The effect size should be gauged *a priori* to the calculations.

5. Alternatively, what would be the width of precision that would be clinically acceptable?

   Not all statistical tests involve hypothesis testing, and sometimes the aim of the study is to measure some quantity. For example, estimate the prevalence of diabetes in a random sample, in which case usually the sample size is calculated so that the confidence interval is within a certain accepted margin (e.g. ±5%).

6. Do you have any estimates of variability in the outcome variable?

   The variance of the outcome is a key ingredient in the sample size calculations, and this needs to be carefully ascertained. For follow-up studies, distinction must be made between variance of follow-up measurement versus variance of change scores. The difference between the standard deviation and the standard error of the mean should also be explained to the collaborator as they are sometimes interchangeably used and interpreted (incorrectly) as being similar.

7. How much is the anticipated dropout or non-completions in your study?

   Accounting of dropouts and non-completions will ensure that the study is adequate powered based on initial calculations. It has been reported among a review of trials published in medical journals that among those trials with some level of loss to follow-up, the median percentage loss was 7% with a minimum of 0.08% to a maximum of 48% (interquartile range 2%–18%) (Dumville et al., 2006).

4.6 Common Mistakes

1. Reporting numbers for only one group/arm

   Most sample size software packages provide estimates for each group/intervention arm that is being compared against another group. For instance, in a two-armed RCT, the software provides numbers that should be recruited in each arm. For the novice, it is easy to mistake this number for the entire study instead, causing the study to be severely under-powered!

2. Not accounting for dropouts

Another common problem in sample size calculations is not accounting adequately for drop-outs or non-completion in the study. This is especially a problem for longitudinal studies with a projected high dropout rate, which can lead to a failure in a study even when correct the sample size technique was performed. A further distinction needs to be made between those who were approached and declined to participate in the study versus those who enrolled in the study and dropped out subsequently. For the former, a comparison of the key demographics and other characteristics would be useful. Sometimes, investigators may provide a cautious estimate for the dropout rate from pilot studies (e.g. 20%). It is equally important for the investigator to design the study in such a way to minimise dropouts (e.g. schedule measurements/questionnaires to be administered during patients' regular visits to clinics, provide transport reimbursement for clinic visits, shorten questionnaire to improve completeness in response, etc.) These ideas can be communicated with the collaborator at the design stage to better address the issue of dropouts.

3. Calculations based on one-sided tests

There is a temptation to calculate the sample size based on one-sided tests as they provide a smaller number, and this should be avoided unless there is a strong justification for one-sided hypothesis testing. For instance, when evaluating the effectiveness of a new drug versus standard treatment, one should not just assume that the new drug will be better but also allow for the possibility of the new drug performing worse.

4. Omitting sample size calculations for selected primary outcome measures

In a study, it is usual to have more than one main outcome measure. For instance, in an RCT evaluating the efficacy of a new corticosteroid drug among asthma patients, the main outcomes could include asthma attacks (exacerbations) or even QoL scores. A clinician may decide to prescribe the drug to his/her patients even if the drug works for one of the two main outcomes, and thus, it is important to make sure the study is adequately powered for both outcomes.

5. Using unjustified estimates of effect sizes in sample size calculation

We have seen how the effect size can affect sample size calculations, and one should thus carefully justify the choice of the effect size in the proposal. Gardner's Effect Size Illustrator (ESI; see Section 4.7)

provides a useful tool that can help understand how selecting different values of the effect size can have an impact on the sample size.

6. Replicating sample size calculations from similar published studies/protocols

   Sometimes, it may be tempting for investigators to use sample size calculations from the published literature looking at a similar intervention/outcome or from clinical trial protocols of drug interventions that are currently under way. One should take note of differences in population characteristics, study setting and any other factors, which may limit the use of such numbers.

7. Using the incorrect standard deviation in follow-up studies

   In a longitudinal or cohort study, interest may lie in the change score (e.g. change in QoL from baseline to 6 months of follow-up). This change score may then be compared between an intervention and a control group. One needs to be careful about using the standard deviation estimates of the change score rather than the score at baseline or follow-up in the sample size calculation (Hulley et al., 2007). Change scores usually have a smaller variance, and one may just end up recruiting more subjects than actually required for the study when the variance of the change scores is not used instead.

4.7 Tools and Resources

1. Free sample size software that can be used for commonly performed analysis in clinical research. A point-and-click software that can be used for studies with dichotomous, continuous, or survival response measures. http://biostat.mc.vanderbilt.edu/wiki/Main/PowerSampleSize. Date accessed: 19 August 2013.

2. Gardner's ESI provides users with an intuitive, clinically informative, interactive online illustrator, which is helpful in determining the apparent importance of the effect size observed between two treatment groups using an advanced graphical approach. http://esi.medicine.dal.ca/what-is-the-esi.html. Date accessed: 19 August 2013.

3. Interpreting Cohen's *d* effect size an interactive visualisation. Created by Kristoffer Magnusson, this website provides an alternative interpretation of Cohen's effect size. According to the probability of superiority and number needed to treat. Interactive graphs and visuals are available. http://rpsychologist.com/d3/cohend/. Date accessed: 27 November 2014.

4. A list of sample size software packages, classified according to the analytical technique and whether they are free or require a fee/costing. http://www.epibiostat.ucsf.edu/biostat/sampsize.html?iframe=tr. Date accessed: 19 August 2013.

5. The following article discusses potential difficulties in sample size calculations, including accounting for lag times for interventions, competing risks and non-proportional hazards in a survival analysis and studies that involve factorial designs (i.e. looking at more than one therapy simultaneously) (Wittes, 2002).

4.8 A Collaborative Case Study

Project title: Tools to improve panic screening in the Emergency Department (TIPS-ED).

Collaborating clinician: Dr. Sharon Sung, PhD; assistant professor, Office of Clinical Sciences, Duke-NUS Graduate Medical School, Singapore.

Study summary: The result of this collaboration was a successful grant application for the Health Services Research – New Investigator Grant from the National Medical Research Council (NMRC) in Singapore. The overall aim of this phase I study was to develop an efficient screening method to detect panic disorder (PD) among emergency department (ED) patients in Singapore.

Aim of this study: The main objective was to compare two brief patient-rated questionnaires (panic module of the Patient Health Questionnaire [PHQ] or the Psychiatric Diagnostic Screening Questionnaire [PDSQ]) and establish which is a more effective screening tool for detecting PD in ED patients with panic-like somatic complaints, based on receiver operating characteristics, sensitivity, specificity and predictive values. The secondary aim involved comparing demographics, clinical characteristics, and service use (i.e. frequency of past-year ED visits and hospital admissions) of participants who do/do not meet criteria for PD.

Sample size calculation: The following is an adaptation of the sample size calculation provided in the grant application:

For the primary aim, 71 PD cases will allow us compare an Area Under the Curve (AUC) of 0.78 versus 0.67 for the PDSQ and PHQ respectively. This calculation is based on our pilot data and the published literature. The total sample size would then be 309, assuming 23% of patients with panic-like somatic complaints meet criteria for PD. For the secondary aim, we estimate that a sample size of 71 participants with PD will allow for 0.80 power to detect an effect size of at least 0.48 using age as a major risk factor. Our projected sample size would also allow prevalence of PD to be estimated at 0.23 within a reasonable 95% confidence interval width of 0.09. Sample size calculation was performed in Sample Size Tables for Clinical Studies V1.0.

Lessons learnt from this collaborative exercise: One main factor of success in this grant application was that the first contact for this project between the principal investigator and the biostatistician was more than 1 year before the actual submission of the grant. This allowed for frequent discussions to be held on the study design and analytical methods between the study teams. More importantly, it was possible to plan and complete a smaller pilot study from which estimates of AUC and prevalence of PD could be ascertained and used in the sample size calculations in the main study. In addition, estimates of dropouts and non-response were calculated from the pilot study and used in the planning of the main study. After looking at the pilot data and discussing about possible power issues in the original primary hypothesis, slight changes were made to the aims and hypothesis.

Key Learning Points

1. Understand the various ingredients in a sample size calculation, and hence plan the most efficient strategy for calculating sample sizes.
2. Realise the importance of effect size in sample size calculation and understand how to effectively elicit appropriate values.
3. Learn how to perform common sample size calculations in clinical research using PSPSS software.
4. Pick up tips on 'questions to ask your collaborator' when asked to work on a sample size calculation.
5. Learn how to avoid common mistakes made in sample size calculations.
6. Remember to engage the collaborator early on during the research to allow time to make changes to the initial proposal/sample size.

References

Batistatou, E., Roberts, C., & Roberts, S. (2014). Sample size and power calculations for trials and quasi-experimental studies with clustering. *Stata J, 14,* 159–175.

Cohen, J. (1988). *Statistical Power Analysis for the Behavioural Sciences* (2nd ed.). London, U.K.: Lawrence Erlbaum Associates.

Cohen, J. (1992). A power primer. *Psychol Bull, 112*(1), 155–159.

Dumville, J. C., Torgerson, D. J., & Hewitt, C. E. (2006). Reporting attrition in randomised controlled trials. *BMJ, 332*(7547), 969–971. doi:10.1136/bmj.332.7547.969.

Dupont, W. D., & Plummer, W. D. (1990). Power and sample size calculations: A review and computer program. *Control Clin Trials, 11,* 116–128.

Earnest, A. (2010). Power of a test. *Proc Singapore Healthcare, 19*(4), 360.

Fleiss, J. L., Cohen, J., & Everitt, B. S. (1969). Large sample standard errors of kappa and weighted kappa. *Psychol Bull, 72,* 323–327. http://biostat.mc.vanderbilt.edu/wiki/Main/PowerSampleSize. Date accessed: 12 August 2013.

Hulley, S. B., Cummings, S. R., Browner, W. S., Grady, D. G., & Newman, T. B. (2007). *Designing Clinical Research*. Philadelphia, PA: Lippincott Williams & Wilkins.

Machin, D., Campbell, M. J., Tan, S. B., & Tan, S. H. (2009). *Sample Size Tables for Clinical Studies* (3rd ed.). West Sussex, U.K.: Wiley-Blackwell.

Moher, D., Hopewell, S., Schulz, K. F., Montori, V., Gotzsche, P. C., Devereaux, P. J., ... Altman, D. G. (2012). CONSORT 2010 explanation and elaboration: Updated guidelines for reporting parallel group randomised trials. *Int J Surg, 10*(1), 28–55. doi:10.1016/j.ijsu.2011.10.001.

Pocock, S. J. (2000). *Clinical Trials – A Practical Approach*. Chichester, U.K.: John Wiley & Sons.

von Elm, E., Altman, D. G., Egger, M., Pocock, S. J., Gotzsche, P. C., & Vandenbroucke, J. P. (2007). The strengthening the reporting of observational studies in epidemiology (STROBE) statement: Guidelines for reporting observational studies. *Lancet, 370*(9596), 1453–1457. doi:10.1016/S0140-6736(07)61602-X.

Wittes, J. (2002). Sample size calculations for randomized controlled trials. *Epidemiol Rev, 24*(1), 39–53.

5

Statistical Analysis Plan

5.1 Introduction

The Statistical Analysis Plan (SAP) is probably the most important document for a collaborating biostatistician to work on, regardless of whether it is for a grant application or a protocol for a project leading to a publication. It is a structured proposal that outlines statistical tests and procedures that would be applied in a project. It often includes the hypothesis to be tested, the statistical test to be conducted, details on how the variables are measured and coded (i.e. the exposure, outcome and confounding variables), the statistical software as well as the level of significance at which the p-values will be evaluated against. Information is also provided here on how any complex data analysis problems are handled (e.g. missing data imputation techniques and adjustment for multiple comparisons).

SAPs are usually found in the methods sections of grant proposals, ethics application forms and clinical trial protocols. After the study is conducted, there may be minor changes to the methodology depending on the nature of data and final sample size, and these are usually explained in the statistical analysis method sections of publications, conferences and other reports.

This chapter provides the biostatistician with guidance on choosing the appropriate statistical test for the study as well as outlines how to execute and interpret some of the more commonly performed statistical tests in the hospital setting using the Stata software. The tests are organised along univariate and multi-variate analyses and include a range of statistical techniques ranging from the independent Student's t-test to autoregressive integrated moving average (ARIMA) models. A unique feature of this chapter is the use of realistic clinical examples to exemplify each of the statistical tests. Stata software is used to derive the output from the analysis, and Stata commands are provided along with the screenshots of the output from the software and an interpretation of the output. Although Stata is used as the exemplar software, the description of the output and ensuing discussion of the methods apply universally to all other software as well. A checklist on choosing the appropriate statistical test for the data based on the objective is also included in this chapter. Key issues to consider in the analysis of data such as missing data and intention to treat analysis are also presented along with 'key questions to ask a clinician' in Section 5.5, which are designed to help the biostatistician better organise and arrive at a final SAP earlier during the collaboration process.

5.2 Choosing the Appropriate Statistical Method

Choosing the appropriate statistical technique to answer a research question is the most commonly performed task of a biostatistician in a clinical setting. This would depend on a number of factors, including obviously the specific hypothesis being tested, the nature of data collected, the distribution of quantitative variable (if applicable), the number of groups being compared, the number of variables being studied and to some extent the sample size. Table 5.1 shows one such guide that a biostatistician may follow in determining the appropriate

TABLE 5.1

Outline of Choice of Common Statistical Tests and Procedures

Univariate Analysis

Type of Outcome Measure	Procedure	
1. Estimate mean	Effect and 95% CI	
2. Estimate proportion	Effect and 95% CI	

Bivariate Analysis

What Would You Like to Do?	Procedure	Non-Parametric Option
1. Compare means		
1.1. Two groups	Independent Student's *t*-test	Mann–Whitney U test
1.2. More than two groups	ANOVA	Kruskal–Wallis test
1.3. Within the same group	Paired *t*-test	Wilcoxon signed-rank test
2. Calculate correlation between two continuous variables	Pearson's correlation coefficient	Spearman's correlation coefficient
3. Compare proportions		
3.1. Two groups	Chi-squared test/Fisher's exact test	
3.2. More than two groups	Chi-squared test	
3.3. Within the same group	McNemar's test	
4. Compare survival	Log-rank test	
5. Measure agreement (continuous variables)	Intra-class correlation coefficient	
6. Measure agreement (categorical variables)	Kappa statistic	
7. Calculate relative risk measures	Immediate commands	

(Continued)

TABLE 5.1 (*Continued*)

Outline of Choice of Common Statistical Tests and Procedures

Multi-Variate Analysis

Type of Outcome Measure	Statistical Model	Repeated Measures Option
1. Continuous	Ordinary least squares regression model	GEE model/mixed model
2. Binary	Binary logistic regression model	GEE model/mixed model
3. Ordinal	Ordinal logistic regression model	GEE model/mixed model
4. Categorical (more than two levels)	Multi-nominal logistic regression model	GEE model/mixed model
5. Count data	Poisson regression model	ARIMA models
6. Survival data	Cox regression model	

statistical technique to use in a data analysis or proposal. The table is split into three different sections – univariate, bivariate and multi-variate – depending on the number of variables involved. For instance, a geriatrician is interested in factors associated with medication non-adherence among dementia patients he/she sees at the clinic. The main outcome, non-adherence, is measured on a dichotomous scale (yes/no), and let's assume that there are five socio-demographic independent variables he/she is interested in. Looking at Table 5.1, we start with the multi-variate section as there are more than two variables involved. As the main outcome measure is binary, we then decide to use the binary logistic regression. A typical paragraph for the statistics section on a grant proposal would read as follows:

> The main outcome measure in this study is medication non-adherence (yes/no). The independent factors being studied include sex, age, years of diagnosis, caregiver support and household family income. The binary logistic regression model would be used to examine factors that are independently and significantly associated with non-adherence to medication. A backward stepwise variable selection method will be used with probability of inclusion and exclusion set at 0.05 and 0.1 respectively. Data analysis will be performed in Stata V13.1 (Stata Corp, College Station, TX, USA).

In another scenario, suppose a health services researcher wants to submit a grant proposal to look at the effectiveness of a multidisciplinary team intervention in terms of reducing the proportion of panic attack cases in 1 month, diagnosed in two emergency departments (one with intervention and the other without). As there are only two variables, and we are comparing proportions across two groups, looking at the bivariate section of Table 5.1, a Chi-squared test would be indicated. The Fisher's exact test would be the alternative test to use if the final sample size was small: usually defined when any of the expected cell count in a 2×2 cross-tabulation has a value less than or equal to 5. Where relevant, the non-parametric equivalent of the tests is also provided. The non-parametric tests make little or no assumptions about the distribution of the data (e.g. normality). For simplicity and ease of application, one may think of just using non-parametric tests for all their data analysis. However, the non-parametric tests are not as powerful or efficient as their parametric equivalent tests, and may require a larger sample size or be less likely to demonstrate statistical significance. A sample analysis plan for the previous example would read something like this:

> The main outcome measure is the proportion of panic attacks diagnosed in the emergency department between two institutions. We propose to use the Chi-squared test to compare the proportion of patients with panic attacks between the institution with and without implementation of multi-disciplinary teams. Data analysis will be performed in Stata V13.1 (Stata Corp, College Station, TX, USA) and level of significance set at 0.05.

It is possible that when the final sample is collected, we end up with a smaller than expected sample size because of dropouts and losses to follow-up. Additionally, in one of the cells, the count may be small (e.g. due to sparse outcome and/or exposure). As a consequence, the investigator may need to use Fisher's exact test in the analysis instead. This slight departure from the planned analysis is acceptable as it is a consequence of the nature of actual data collected. Similar instances occur when we have continuous variables that are not normally distributed even after transformation. This may require the use of non-parametric methods in the analysis.

5.3 Common Statistical Hypotheses and Tests

The basic output from a statistical test is the p-value, and it is defined as the probability of observing the sample or more extreme values of the sample when the null hypothesis is true. Another interpretation of the p-value is the probability of observing the results due to chance. Consequently, if this value is very low, we conclude that the results from the experiment are not due to chance and are real. Often, the threshold that is used is 0.05, also known as the level of significance, and this implies that there is a 1 in 20 chance of making a type 1 error, that is, the probability of rejecting the null hypothesis and claiming a significant result when in fact there isn't. In this section, using a series of clinical scenarios, we demonstrate how to perform some of the more common statistical techniques in clinical research using Stata and interpret the output. The focus is not on the theoretical side of the tests, and thus, the display of mathematical formulae and equations will be minimised. Those interested in delving into the mathematics behind the tests are referred to the many other biostatistics textbooks, including 'Statistical methods in Medical Research' by Armitage and colleagues (2002), which provide the details. The outline in Table 5.1 will also be used as a guide in this chapter.

5.3.1 Univariate Analysis

5.3.1.1 Estimate Mean

An internal medicine specialist wanted to calculate and report the mean body mass index (BMI) level for his/her sample of 300 patients along with the corresponding 95% confidence interval (CI). As we can see from Figure 5.1, typing in the command 'ci [varname]' in Stata will provide us the relevant values. Alternatively, if we only had the summarised value of the mean and standard deviation (SD) available instead of the raw data, then Stata also allows for an immediate form of the command: cii #obs #mean #sd. As we can see from the bottom of Figure 5.1, we get the same answer. The data shows that the mean BMI among patients was 28.2 kg/m^2 (95% CI: 27.4–29.0).

5.3.1.2 Estimate Proportion

Now, assume the same clinician wanted to estimate the proportion of patients who currently smoke. In Stata, we type 'ci [varname], binomial'. As we can see from Figure 5.2, Stata provides the proportion of smokers as 47.3% with a 95% CI of 41.6–53.1. Similarly, if we only had aggregate data, we type 'cii #obs #succ or 'cii 300 142', which provides the same results (Figure 5.2).

```
ci bmi
```

Variable	Obs	Mean	Std. Err.	[95% Conf. Interval]	
bmi	300	28.22	.4203502	27.39278	29.04722

```
su bmi
```

Variable	Obs	Mean	Std. Dev.	Min	Max
bmi	300	28.22	7.28068	17	40

```
cii 300 28.22 7.28
```

Variable	Obs	Mean	Std. Err.	[95% Conf. Interval]	
	300	28.22	.420311	27.39286	29.04714

Immediate form of the command. This is helpful when only summarised data are available.

FIGURE 5.1
Calculating mean and 95% CI for a continuous variable.

```
ci smoking, binomial
```

Variable	Obs	Mean	Std. Err.	— Binomial Exact — [95% Conf. Interval]	
smoking	300	.4733333	.0288264	.4156768	.5315216

```
tab smoking
```

Smoking	Freq.	Percent	Cum.
0	158	52.67	52.67
1	142	47.33	100.00
Total	300	100.00	

```
cii 300 142
```

Variable	Obs	Mean	Std. Err.	— Binomial Exact — [95% Conf. Interval]	
	300	.4733333	.0288264	.4156768	.5315216

FIGURE 5.2
Calculating proportion and 95% CI.

5.3.2 Bivariate Analysis

5.3.2.1 Compare Means in Two Groups

A hospital administrator wanted to examine a specific hypothesis that females stayed longer than males in the hospital. To specify an independent Student's t-test in Stata, type 'ttest [varname], by([groupvar])' or 'ttest los, by(gender), and the output is shown in Figure 5.3. As hypothesised, females indeed stayed on average 2.8 days (95% CI: 2.2–3.5)

```
. ttest los, by(gender)
```

Two-sample t test with equal variances

Group	Obs	Mean	Std. Err.	Std. Dev.	[95% Conf. Interval]	
Male	156	7.187831	.2460178	3.072761	6.70185	7.673811
Female	144	10.00381	.2218602	2.662323	9.565263	10.44236
combined	300	8.539502	.185026	3.204745	8.175384	8.90362
diff		-2.815982	.3331792		-3.471664	-2.1603

```
    diff = mean(Male) - mean(Female)                           t =  -8.4519
Ho: diff = 0                                 degrees of freedom =      298

    Ha: diff < 0                 Ha: diff != 0                  Ha: diff > 0
Pr(T < t) = 0.0000       Pr(|T| > |t|) = 0.0000          Pr(T > t) = 1.0000
```

Two-sided p-value

FIGURE 5.3
Independent student's t-test.

longer than males. Note that the *p*-value should be obtained from the two-sided test (indicated by arrow) and should be quoted as <.001 and not 0!

5.3.2.2 Assumptions of Independent Student's t-Test

1. *Independence of observations.* This assumption is fulfilled as we know from the design of the study that each of the 300 patients is unique and not duplicated in the sample.

2. *Homogeneity of variance.* The variance appears slightly higher among males (SD = 3.1) versus females (SD = 2.7), although a formal test indicates that they are not significantly different ($p = .082$) (Figure 5.4). This is obtained by typing 'sdtest [varname], by[groupvar]' or in our example 'sdtest los, by(gender)' in Stata.

3. *Normality of distribution.* From the quantile-normal plot (bottom left) in Figure 5.4, we can observe that the distribution for length of

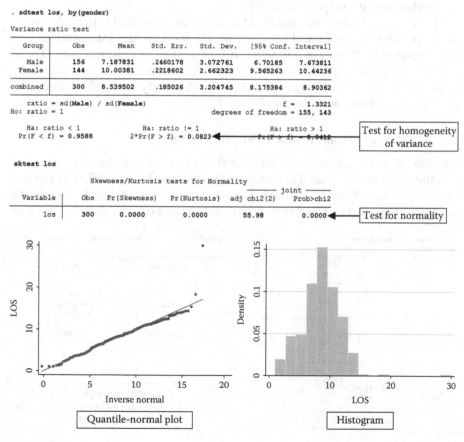

FIGURE 5.4
Testing assumptions for the independent Student's *t*-test.

stay (LOS) is positively skewed. From the histogram, it is obvious that there is an outlier (patient staying around 30 days) that is causing the data to be skewed. A more formal test for departure from normality can be obtained by typing 'sktest los', and as we can see from Figure 5.4, the *p*-value is <.001, indicating that we should reject the null hypothesis that the data come from a normal distribution.

5.3.2.3 Non-Parametric Equivalent: Mann–Whitney U Test

When the assumption of normality is violated as we saw in the previous example, and if transformation of the data does not help, one may wish to use a non-parametric test such as the Mann–Whitney U test (also known as the Wilcoxon rank-sum test). This test does not make any assumption about the distribution of the variable and is instead based on the rank of the values. In Stata, type 'ranksum los, by(gender)'. As shown in Figure 5.5, the sum of ranks is compared across males and females with a corresponding significant *p*-value (<.001).

Tip: When we have conceded that the data are skewed, it is probably wise not to present the mean and SD. Instead, the median and interquartile can be calculated as these are not affected by the skewness in the data. These can be easily obtained in Stata by typing 'tabstat los, by(gender) stats(median p25 p75)', with p25 and p75 signifying the 25th and 75th percentiles of los, respectively (Figure 5.5).

5.3.2.4 Compare Means in More Than Two Groups

Suppose now the same investigator wanted to examine the hypothesis of whether patients in older age groups stayed longer in hospital, and the following are the three age groups he/she was interested in: (1) <40 years, (2) 40–64 years and (3) 65+ years. From Table 5.1, we can see that the analysis

```
. ranksum los, by(gender)

Two-sample Wilcoxon rank-sum (Mann-Whitney) test

    gender |     obs    rank sum    expected
-----------+--------------------------------
      Male |     156       17443       23478
    Female |     144       27707       21672
-----------+--------------------------------
  combined |     300       45150       45150

unadjusted variance      563472.00
adjustment for ties          -0.50
                        ----------
adjusted variance        563471.50

Ho: los(gender==Male) = los(gender==Female)
              z =  -8.040
    Prob > |z| =   0.0000
```

Mann–Whitney U test

```
. tabstat los, by(gender) stats(median p25 p75)

Summary for variables: los
    by categories of: gender (Gender)

 gender |      p50         p25         p75
--------+------------------------------------
   Male | 7.039634    4.813759     9.24486
 Female | 9.569317    8.403561    11.21688
--------+------------------------------------
  Total | 8.503399    6.456333    10.68656
```

Calculating descriptive statistics for data with a skewed distribution

FIGURE 5.5
Results from a Mann–Whitney U test.

of variance (ANOVA) test would be the appropriate one to use. In Stata, type 'anova los agegroup', and we would obtain the output shown in Figure 5.6. The ANOVA test is based on the sum of squares, and the mean squared error for the model is compared with that of the error term, giving rise to the *F*-statistic. As we can see from the *p*-value of .382, we do not have evidence to reject the null hypothesis that the mean LOS in the three age groups is equal. The tabstat command can be used to calculate and display the means and SDs in the three groups as shown here: tabstat los, by(agegroup) stats(mean sd). The mean LOS was 8.2 for patients aged less than 40 years, 8.8 for patients aged 40–64 and 8.7 for those aged 65 and above, but these differences were not statistically significant ($p = .382$).

5.3.2.5 Assumptions of the ANOVA Test

The assumptions for the ANOVA test are similar to the *t*-test:

1. Independence of observations
2. Homogeneity of variance
3. Normality of distribution

```
anova los agegroup
```

	Number of obs =	300	R-squared	= 0.0065
	Root MSE	= 3.20512	Adj R-squared	= -0.0002

Source	Partial SS	df	MS	F	Prob > F
Model	19.8262414	2	9.91312068	0.96	0.3822
agegroup	19.8262414	2	9.91312068	0.96	0.3822
Residual	3051.02069	297	10.2727969		
Total	3070.84693	299	10.2703911		

ANOVA test

```
. tabstat los, by(agegroup) stats(mean sd)
```

Summary for variables: los
    by categories of: agegroup

agegroup	mean	sd
less than 40	8.169279	2.78243
40-64	8.763464	3.244542
65 and above	8.666827	3.513555
Total	8.539502	3.204745

Descriptive statistics

FIGURE 5.6
Results from an ANOVA test.

These assumptions will be discussed in more details in Section 5.3.3.2.

5.3.2.6 Non-Parametric Equivalent: Kruskal–Wallis Test

In the event that the distributional assumptions in an ANOVA test are not fulfilled, one may wish to perform the non-parametric equivalent Kruskal–Wallis test instead. In Stata, type 'kwallis los, by(agegroup)'. As we can see from Figure 5.7, the *p*-value is .339, indicating that the results are still not statistically significant. Note also how the sum of ranks is provided for each group, which goes towards the calculation of the *p*-value. Before choosing any non-parametric test, one may wish to try some transformation to see if the data can be made normally distributed. In Stata, there is a command that provides for the quantile-normal plots after various power transformations. The Stata command qladder los would provide us with various plots as shown in Figure 5.8. It seems that the logarithmic transformation would be the most appropriate in terms of bringing the points towards the line. Sometimes, in the presence of extreme skewness, the identity (equivalent to no transformation) would indicate the best fit, in which case a non-parametric equivalent model may be more appropriate. The following command in Stata can be used to create a new log-transformed variable for los: gen log_los=log(los); subsequently a parametric test can be performed on this transformed variable.

5.3.2.7 Compare Means within the Same Group

An endocrinologist wanted to compare low-density lipoprotein (LDL) levels among his/her diabetic patients at first visit (LDL1) versus 1 year later (LDL2). He/she hypothesised that there would be a change in the mean values over this period of disease instability. Because we are comparing

```
. kwallis los, by( agegroup)

Kruskal-Wallis equality-of-populations rank test
```

agegroup	Obs	Rank Sum
less than 40	96	13521.00
40-64	99	15747.00
65 and above	105	15882.00

```
chi-squared =       2.162 with 2 d.f.
probability =       0.3393

chi-squared with ties =       2.162 with 2 d.f.
probability =       0.3393
```

Kruskal–Wallis test

FIGURE 5.7
Results from the Kruskal–Wallis test.

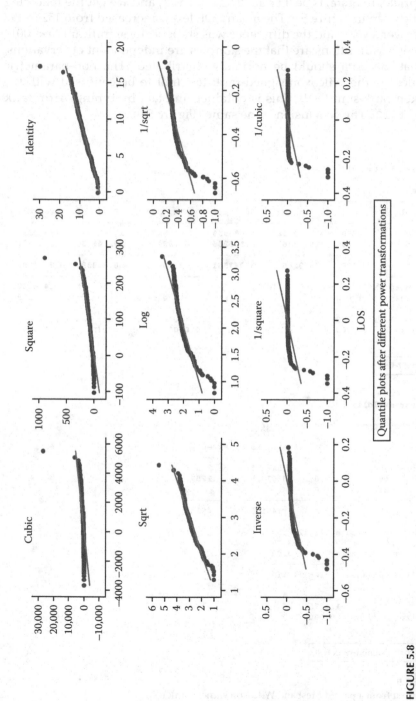

FIGURE 5.8
Quantile-normal plots for various transformations of LOS.

the mean values of LDL at two different time points, a paired *t*-test would be appropriate. In Stata, type 'ttest ldl1= ldl2', and we get the following output shown in Figure 5.9. The mean LDL levels increased from 152 to 197 mg/dL over a year, and the difference was statistically significant ($p < .001$). The main assumptions are that the samples are independent observations, and that the data should be normally distributed. The non-parametric equivalent is the Wilcoxon signed-rank test (not to be confused with the Wilcoxon rank-sum test!). This is obtained in Stata by typing 'signrank ldl1= ldl2'. The conclusion is the same (Figure 5.9).

. **ttest ldl1= ldl2**

Paired t test

Variable	Obs	Mean	Std. Err.	Std. Dev.	[95% Conf. Interval]	
ldl1	300	152.52	1.686959	29.21898	149.2002	155.8398
ldl2	300	197.3467	2.505356	43.39405	192.4163	202.277
diff	300	-44.82667	3.067551	53.13153	-50.86339	-38.78994

```
     mean(diff) = mean(ldl1 - ldl2)                              t = -14.6132
Ho: mean(diff) = 0                          degrees of freedom =       299

Ha: mean(diff) < 0           Ha: mean(diff) != 0           Ha: mean(diff) > 0
Pr(T < t) - 0.0000         Pr(|T| > |t|) - 0.0000         Pr(T > t) - 1.0000
```

Paired *t*-test

. **signrank ldl1= ldl2**

Wilcoxon signed-rank test

sign	obs	sum ranks	expected
positive	72	5470.5	22575
negative	228	39679.5	22575
zero	0	0	0
all	300	45150	45150

```
unadjusted variance   2261262.50
adjustment for ties       -93.38
adjustment for zeros        0.00
                      ----------
adjusted variance     2261169.13

Ho: ldl1 = ldl2
            z = -11.375
   Prob > |z| =    0.0000
```

Wilcoxon signed-rank test

FIGURE 5.9
Stata output from a paired *t*-test and Wilcoxon signed-rank test.

5.3.2.8 Calculate Correlation between Two Continuous Variables

A rheumatologist at a tertiary hospital was interested in looking at the association between age and quality of life (QoL), both of which are continuous variables, and QoL measured on a scale from 0 to 100. Based on Table 5.1, we need to calculate the Pearson correlation coefficient. In Stata, we type 'pwcorr age qol, obs sig'. In the Stata output (Figure 5.10), the Pearson correlation coefficient is −0.06 with a *p*-value of .298. It appears that there is a negative weak linear relationship between age and QoL. The non-parametric equivalent is the Spearman correlation coefficient, and in Stata, the corresponding command is 'spearman age qol'.

In theory, the correlation coefficient ranges from −1 (perfect negative linear association) to +1 (perfect positive linear association) with values closer to 0 indicating no correlation. One should be cautious about the interpretation of the correlation coefficient. For instance, correlation does not imply causation. In our earlier example, if we found a strong positive correlation between age and QoL, it does not mean that old age causes one to have a poor QoL, or conversely people with poorer QoL get older! The other main point to note is that the correlation quantifies a linear relationship. In some studies, the associations between variables can be non-linear. For example, Figure 5.11 shows a u-shaped relationship between age and muscle strength with maximal strength observed in the age group 30–40. One should then be cautious about calculating and interpreting the correlation coefficient in such instances.

```
. pwcorr age qol, obs sig
```
                                                              Pearson correlation coefficient
```
                  |     age        qol

          age |  1.0000

              |    300

          qol | -0.0603     1.0000
              |  0.2975
              |    300        300
```
                                                              Note: Stata provides the correlation
                                                              coefficient in the first row of
                                                              numbers, *p*-value in the second
                                                              and sample size in the third.

```
. spearman age qol

  Number of obs =       300
  Spearman's rho =    -0.0728

  Test of Ho: age and qol are independent
       Prob > |t| =       0.2085
```
                                                              Spearman's correlation coefficient

FIGURE 5.10
Stata output for Pearson's and Spearman's correlation coefficients.

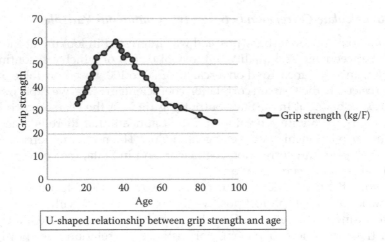

U-shaped relationship between grip strength and age

FIGURE 5.11
Non-linear relationship between age and muscle strength.

5.3.2.9 Compare Proportions

5.3.2.9.1 In Two Groups

Suppose a respirologist wanted to compare the proportion of deaths at 6 months between smokers and non-smokers. From Table 5.1, we can see that a Chi-squared test or Fisher's exact test may be appropriate. In Stata, we type 'tab dead _ 6mth smoking, col chi expected'. This produces a cross-tabulation between death at 6 months and current smoking status along with the expected cell count, and the results from the Chi-squared test (Figure 5.12). This test will suffice as none of the expected cell counts are less than 5. Otherwise, a Fisher's exact test may be required. In that case, we then substitute with the following command in Stata: tab dead _ 6mth smoking, col exact expected. As we can see, the proportion of deaths among smokers and non-smokers is similar (53.5% vs. 54.4%, respectively), and the result is not statistically significant ($p = .875$).

5.3.2.9.2 In More Than Two Groups

Suppose now the respirologist wanted to compare the proportion of deaths across the following three age groups: (1) <40 years, (2) 40–64 years, and (3) 65 years and above. Notice now that we can now swop the axes and request for a row percentage instead and leave out the expected count, with the following command in Stata: 'tab agegroup dead_6mth, row chi expected'. As we can see from Figure 5.13, the proportion of deaths is higher among those aged <40 years (65%) versus 49% in each of the other two age groups, and the corresponding p-value from the Chi-squared test is significant ($p = .041$). Note that the p-value indicates that any one of the pairs of categories are statistically significant, but we may not necessarily know which ones, unless we further classify the information into 2 × 2 tables and run the test again.

```
. tab dead_6mth smoking, col chi expected
```

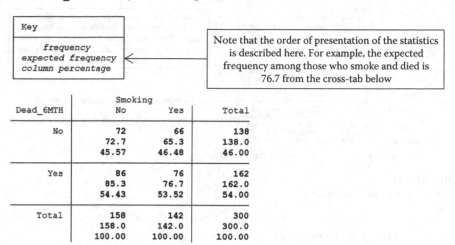

Key
frequency
expected frequency
column percentage

Note that the order of presentation of the statistics is described here. For example, the expected frequency among those who smoke and died is 76.7 from the cross-tab below

	Smoking		
Dead_6MTH	No	Yes	Total
No	72	66	138
	72.7	65.3	138.0
	45.57	46.48	46.00
Yes	86	76	162
	85.3	76.7	162.0
	54.43	53.52	54.00
Total	158	142	300
	158.0	142.0	300.0
	100.00	100.00	100.00

Pearson chi2(1) = 0.0249   Pr = 0.875

FIGURE 5.12
Stata output from a Chi-squared test (smoking with death).

```
. tab agegroup dead_6mth , row chi expected
```

Key
frequency
expected frequency
row percentage

	Dead_6MTH		
agegroup	No	Yes	Total
less than 40	34	62	96
	35.42	51.8	96.0
	35.42	64.58	100.00
40-64	50	49	99
	45.5	53.5	99.0
	50.51	49.49	100.00
65 and above	54	51	105
	48.3	56.7	105.0
	51.43	48.57	100.00
Total	138	162	300
	138.0	162.0	300.0
	46.00	54.00	100.00

Pearson chi2(2) = 6.3833   Pr = 0.041

Chi-squared test (more than two categories)

FIGURE 5.13
Stata output from a Chi-squared test (age group with death).

```
. mcc joint1 joint2
```

Cases	Controls Exposed	Unexposed	Total
Exposed	58	69	127
Unexposed	66	56	122
Total	124	125	249

> *Note:* mcc denotes matched case control, and by default, Stata treats the first variable listed as case, and the next control. In our example, joint 1 is joint pain before treatment, and joint 2 is joint pain after treatment.

```
McNemar's chi2(1) =      0.07    Prob > chi2 = 0.7963
Exact McNemar significance probability       = 0.8634
```

```
Proportion with factor
     Cases        .5100402 ⟵────────────────────────── Proportion of joint pain before treatment
     Controls     .497992  ⟵── [95% Conf. Interval]
                                                     ── Proportion of joint pain after treatment
     difference   .0120482    -.0834123   .1075087
     ratio        1.024194     .8542226   1.227985
     rel. diff.   .024        -.1559823   .2039823

     odds ratio   1.045455     .7351042   1.487924   (exact)
```

FIGURE 5.14
Stata output from a McNemar test.

5.3.2.9.3 *Within the Same Group*

A rheumatologist evaluated the presence/absence of joint pain among 250 of his/her patients before and after 3 months of starting treatment. He/ she wanted to compare the proportion of joint pain at the two time points. From Table 5.1, we observe that a McNemar test would be useful. In Stata, type 'mcc joint1 joint2'. From Figure 5.14, we see that the proportions of joint pain did not change significantly before (51%) and after treatment (49%), $p = .796$.

5.3.2.10 *Compare Survival*

A cardiologist wanted to compare the survival between patients with coronary heart disease (CHD) and those without CHD. In Stata, firstly we need to set the data up for a survival (or time to an event) analysis, including specifying the follow-up and failure variables. In Stata, we type 'stset followupdays, failure(died)'. From the Stata output in Figure 5.15, we see that there are 300 observations with 196 deaths, and the longest follow-up period was for 716 days. Before performing the actual test, it is a good idea to visually examine the survival curves in the two groups. This is achieved with the following command: sts gr, by(CHD) ploto(lpattern(-)) plot2(lpattern(1)). It appears that the survival is lower among those with CHD. Next, to perform the log-rank test, type 'sts test CHD'. As we would expect, survival is indeed lower among those with CHD compared to those without, and the log-rank test indicates that this is statistically significant ($p < .001$) (Figure 5.15).

5.3.2.11 *Measure Agreement (Continuous Variables)*

A nurse practitioner wanted to examine the level of agreement of a new pain score tool that has just been created. The scores in the questionnaire ranged from 0 to 100, and the tool was administered to 10 patients on two

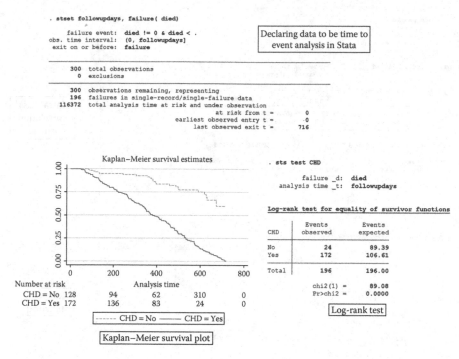

FIGURE 5.15
Stata output for a survival analysis.

consecutive days. From Table 5.1, it is determined that an intra-class correlation coefficient is needed. In Stata, we type 'icc pain pat _ id'. Note that the data should be in a 'long' format, rather than the usual 'wide' format (Figure 5.16). The correlation of pain score for the same individual is 0.58. The correlation of mean score across both days is 0.73. According to thresholds (Table 5.2) (Landis & Koch, 1977) developed for kappa statistic (also reasonable to be used here), there is 'moderate' level of agreement.

5.3.2.12 Measure Agreement (Categorical Variables)

A senior consultant in a radiology department wanted to look at the level of agreement between him/her and his/her registrar in terms of detecting brain tumours (yes/no) from a series of 15 magnetic resonance imaging images. In Stata, we type the command: kap radiologist1 radiologist2, tab. As we can see from Figure 5.16 (bottom test), the level of agreement between the senior consultant and the registrar is given by $\kappa = 0.44$, with $p = .043$. According to Landis and Koch's guidelines in Table 5.2 (Landis & Koch, 1977), the kappa value of 0.44 can be characterised as again having moderate level of agreement, and this is statistically significant from a value of 0.

```
. icc pain pat_id

Intraclass correlations
One-way random-effects model
Absolute agreement

Random effects: pat_id           Number of targets =      10
                                 Number of raters  =       2

                pain  |   ICC      [95% Conf. Interval]
          Individual  | .576007   -.0082581   .8728861
             Average  | .7309701  -.0166538   .9321294

F test that
   ICC=0.00: F(9.0, 10.0) = 3.72          Prob > F = 0.026

Note: ICCs estimate correlations between individual measurements
      and between average measurements made on the same target.
```

pat_id	day	pain
1	1	18
1	2	20
2	1	36
2	2	39
3	1	34
3	2	41
4	1	34
4	2	84
5	1	71
5	2	75
6	1	76
6	2	85
7	1	78
7	2	14
8	1	98
8	2	98
9	1	60
9	2	52
10	1	27
10	2	22

Intra-class correlation coefficient

```
. kap radiologist1 radiologist2, tab

Radiologis |       Radiologist2
        t1 |     0          1   |   Total
         0 |     4          2   |     6
         1 |     2          7   |     9
     Total |     6          9   |    15

              Expected
Agreement   Agreement   Kappa   Std. Err.      Z    Prob>Z
  73.33%      52.00%    0.4444   0.2582      1.72    0.0426
```

scan_id	radiologist1	radiologist2
1	1	1
2	0	0
3	1	1
4	1	1
5	1	0
6	1	0
7	1	1
8	1	1
9	1	1
10	1	1
11	0	1
12	0	0
13	0	1
14	0	0
15	0	0

Kappa statistic

Measures of agreement

FIGURE 5.16
Stata output for agreement studies.

TABLE 5.2

Cut-offs for Kappa Values and Strength of Agreement

Kappa	Strength
<0	Poor
0–0.20	Slight
0.21–0.40	Fair
0.41–0.60	Moderate
0.61–0.80	Substantial
0.81–1	Almost perfect

Source: Landis, J.R. & Koch, G.G., *Biometrics*, 33, 159–74, 1977. With permission.
Note: Cut-off values obtained from the work of Landis and Koch.

5.3.2.13 Relative Risk Measures

Rate ratios, risk ratios and odds ratios (ORs) are all common measures of effect size in epidemiological studies. Here, we look at some examples of formulating the commands in Stata and interpreting them. Suppose a hospital epidemiologist followed up 20 patients in a geriatric ward for 10 days and another 20 patients in a general ward for 20 days, and calculated the number of new methicillin-resistant *Staphylococcus aureus* (MRSA) infection cases to be 7 and 6, respectively. Because the length of hospital stay is going to affect the risk of MRSA, we need to standardise by using person-days in the denominator during the comparison. To calculate the incidence rate ratio (IRR), type 'iri 7 6 200 400' in Stata. As we can see from Figure 5.17, the IRR for MRSA for the geriatric ward was 2.33 (95% CI: 0.67–8.40) compared to the general ward, but the *p*-value was not significant (*p* = .138).

```
iri 7 6 200 400
```

	Exposed	Unexposed	Total
Cases	7	6	13
Person-time	200	400	600
Incidence rate	.035	.015	.0216667
	Point estimate		[95% Conf. Interval]
Inc. rate diff.		.02	-.0085711	.0485711
Inc. rate ratio		2.333333	.6714592	8.404071 (exact)
Attr. frac. ex.		.5714286	-.4892938	.8810101 (exact)
Attr. frac. pop		.3076923		

(midp) Pr(k>=7) =		0.0691 (exact)
(midp) 2*Pr(k>=7) =		0.1382 (exact)

Rate ratio

```
. csi 4 90 100 100
```

	Exposed	Unexposed	Total
Cases	4	90	94
Noncases	100	100	200
Total	104	190	294
Risk	.0384615	.4736842	.3197279
	Point estimate		[95% Conf. Interval]
Risk difference		-.4352227	-.5152637	-.3551816
Risk ratio		.0811966	.0307012	.2147432
Prev. frac. ex.		.9188034	.7852568	.9692988
Prev. frac. pop		.3250189		

chi2(1) =	58.53	Pr>chi2 = 0.0000

Risk ratio

```
. cci 45 20 55 10
```

	Exposed	Unexposed	Total	Proportion Exposed
Cases	45	20	65	0.6923
Controls	55	10	65	0.8462
Total	100	30	130	0.7692
	Point estimate		[95% Conf. Interval]	
Odds ratio	.4090909		.1553289	1.033738 (exact)
Prev. frac. ex.	.5909091		-.0337383	.8446711 (exact)
Prev. frac. pop	.5			

chi2(1) =	4.33	Pr>chi2 = 0.0374

Odds ratio

Epidemiological measures of relative risk

FIGURE 5.17
Stata output for various measures of relative risk.

The same epidemiologist then surveyed 104 geriatric patients and another different 190 general ward patients to find out how many had been diagnosed with diabetes and found that to be 4 and 90, respectively. To calculate the relative risk ratio (RRR), type 'csi 4 90 100 100' in Stata. Note that these four numbers represent exposed and unexposed cases and exposed and unexposed controls, respectively. Figure 5.17 shows that the RRR of diabetes among those in geriatric ward was 0.08 (95% CI: 0.03–0.21), and this was statistically significant ($p < .001$).

Suppose in another case–control study, a medical oncologist wanted to compare the presence of mutation in a new genetic marker BRCA3 among 65 breast cancer patients and an equal number of controls in his/her clinic. Test results indicated that 45 and 55 patients in the case and control groups showed mutation in the genes. To calculate the OR, type 'cci 45 20 55 10' in Stata. As shown in Figure 5.17, the OR for having a positive test result (i.e. mutation in *BRCA3* genes) among the breast cancer patients was 0.41 (95% CI: 0.16–1.03).

5.3.3 Multi-Variate Analysis

5.3.3.1 Continuous Outcome Measure–Linear Regression Model

An endocrinologist wanted to examine whether factors such as gender, age and presence of diabetes were associated with LDL levels among his/her patients. Because LDL is a continuous outcome measure, we use the ordinary least squares linear regression model (Table 5.1) to analyse the data. The explanatory variables can be continuous or categorical.

Firstly, we look at the association between each of the explanatory variables with LDL levels (univariate analysis). In Stata, we type 'xi: regress ldl1 i.gender'. The xi prefix is to inform Stata that there may follow a categorical explanatory variable for which we need to put an i. prefix in front of the explanatory variable. Otherwise, Stata will treat any explanatory variable as continuous in nature. We repeat the command for all the other factors. As we can see from Figure 5.18, only gender and diabetes are significant in the univariate. The mean LDL value among females (coded 2) is 7.6 mg/dL higher than among males, and the results are statistically significant ($p = .024$). The adjusted R-squared model shows that 1% of the variation in LDL levels is explained by gender. In the third output, the presence of diabetes is also significant, with mean difference of 20.8 mg/dL compared to non-diabetics.

There are several strategies to model building for the multi-variate analysis. One may wish to throw in all variables (regardless of whether significant or not in the univariate analysis) in the multi-variate selection model, but this may not be a wise move, especially if there are lots of missing observations for the variables. In Stata, there are several 'automatic' model selection techniques (type 'help stepwise' to see the choices), but the results usually do not vary, although one needs to explicitly state the method used

```
. xi: regress ldl1 i.gender
i.gender          _Igender_1-2          (naturally coded; _Igender_1 omitted)
```

Source	SS	df	MS	
Model	4356.01034	1	4356.01034	
Residual	250914.87	298	841.996207	
Total	255270.88	299	853.748763	

Number of obs = 300
F(1, 298) = 5.17
Prob > F = 0.0236
R-squared = 0.0171
Adj R-squared = 0.0138
Root MSE = 29.017

ldl1	Coef.	Std. Err.	t	P>|t|	[95% Conf. Interval]	
_Igender_2	7.627137	3.353298	2.27	0.024	1.027992	14.22628
_cons	148.859	2.323233	64.07	0.000	144.287	153.431

Univariate analysis

ldl1	Coef.	Std. Err.	t	P>|t|	[95% Conf. Interval]	
age	-.0548456	.0697325	-0.79	0.432	-.1920761	.0823848
_cons	155.4612	4.102861	37.89	0.000	147.3869	163.5354

ldl1	Coef.	Std. Err.	t	P>|t|	[95% Conf. Interval]	
_Idiabetes_1	20.75355	3.85282	5.39	0.000	13.17137	28.33573
_cons	136.4706	3.388146	40.28	0.000	129.8029	143.1383

```
. xi:stepwise, pr(.2) pe(0.05): regress ldl1 i.gender i.diabetes
i.gender          _Igender_1-2          (naturally coded; _Igender_1 omitted)
i.diabetes        _Idiabetes_0-1        (naturally coded; _Idiabetes_0 omitted)
                  begin with full model
p < 0.2000        for all terms in model
```

Source	SS	df	MS	
Model	26805.7813	2	13402.8906	
Residual	228465.099	297	769.242757	
Total	255270.88	299	853.748763	

Number of obs = 300
F(2, 297) = 17.42
Prob > F = 0.0000
R-squared = 0.1050
Adj R-squared = 0.0990
Root MSE = 27.735

Multi-variate analysis

ldl1	Coef.	Std. Err.	t	P>|t|	[95% Conf. Interval]	
_Igender_2	7.450531	3.20532	2.32	0.021	1.142514	13.75855
_Idiabetes_1	20.66287	3.824868	5.40	0.000	13.1356	28.19015
_cons	132.9645	3.686139	36.07	0.000	125.7102	140.2187

Ordinary least-squares linear regression model

FIGURE 5.18
Stata output for a linear regression model.

in the methods section to ensure reproducibility of results. There are some problems with stepwise selection, and these are discussed in the Stata forum list at http://www.stata.com/support/faqs/statistics/stepwise-regression-problems/ (Date accessed: 5 December 2014). For the multi-variate model, we type the following command in Stata: xi:stepwise, pr(.2) pe(0.05): regress ldl1 i.gender i.diabetes, which allows for the probability of removal and entry of variables into the model to be set at .20 and .05, respectively. As shown in Figure 5.18, both gender and diabetes are still independently and significantly associated with LDL levels, although there are slight changes to the magnitude of the effect size. The adjusted R-squared shows that approximately 10% of the variation in LDL values can be explained by these two variables.

5.3.3.2 Assumptions of the Linear Regression Model

1. *Independence of observations.* Samples should come from independent patients.
2. *Homogeneity of variance.* The error variance should be constant.
3. *Normality of error terms.* The error terms should follow a normal distribution.
4. *Independent variables are not multicollinear.* The variables are not near perfect linear combinations of one another.

The first assumption of independence is usually ascertained in the study design stage where we ensure that LDL measurements are not taken more than once from the same patient. For the homogeneity of variance of error term, we type 'rvfplot, yline(0)' in Stata to visually examine the plot. In Figure 5.19, we can see that the residuals are fairly equally spread across the range of predicted values of LDL. To examine whether the residuals (error terms) are normally distributed, firstly we calculate the residuals using the following command predict r, residuals. Then we plot the quantile-normal figure with 'qnorm r' and finally perform a formal test 'sktest r'. From Figure 5.19, we can see that that there is no obvious departure from normality. Finally, to assess collinearity between the explanatory/independent variables, we type 'vif' in Stata. Variance inflation factor (VIF) values above 10 usually indicate multicollinearity, or equivalently tolerance (1/VIF) values less than 0.1 can also be used. In our example, this assumption is not violated. In the event of multicollinearity, one may wish to keep the variable that is most strongly associated with the outcome and leave the other one out of the final multi-variate model.

   In addition to the above assumptions, one should also be careful not to extend the regression line or relationship beyond the values from which the data was derived from. For instance, if a study included those above 16 years and determined a linear relationship between intelligence quotient (IQ) and age, one cannot simply assume the same linear relationship holds for those who are younger than 16.

5.3.3.3 Binary Outcome Measure–Binary Logistic Regression Model

A rehabilitation medicine specialist planned to find out if demographic and clinical variables could help explain the 6-month survival (yes/ no) of stroke patients who were undergoing rehabilitation treatment at the Alfred Centre. In particular, he/she was interested in the following explanatory variables: age group, number of comorbidities and patient's self-reported QoL. As the

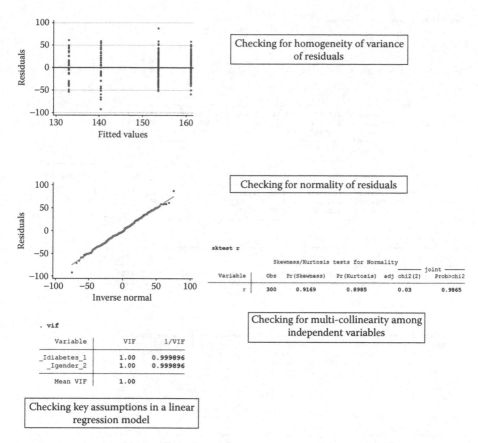

FIGURE 5.19
Checking for assumptions in a linear regression model.

primary outcome is binary in nature, from Table 5.1 we see that a binary logistic regression model is appropriate. The explanatory variables can be continuous or categorical. Similar to the linear regression model described in Section 5.3.3.1, we run the univariate analysis followed by a multi-variate stepwise regression.

In Stata, we type 'xi:logistic survive i.agegroup'. From Figure 5.20, we can see that age group is significantly associated with survival at 6 months. The OR of survival among those aged 40–64 years was 0.4 (95% CI: 0.2–0.8) compared to those less than 40 years. Those aged 65 and above were also less likely to survive (OR = 0.56), although the p-value was not significant ($p = .061$). Similarly, we find that the number of comorbidities and QoL were also significantly associated with survival. For QoL, it appears that the OR of survival increased by a factor of 1.01 for every unit increase in QoL. Note that the interpretation for categorical explanatory variables will be different from continuous

```
. xi:logistic survive i.agegroup
i.agegroup      _Iagegroup_0-2    (naturally coded; _Iagegroup_0 omitted)

Logistic regression                          Number of obs   =        300
                                             LR chi2(2)      =       8.52
                                             Prob > chi2     =     0.0141
Log likelihood = -189.34624                  Pseudo R2       =     0.0220
```

survive	Odds Ratio	Std. Err.	z	P>|z|	[95% Conf. Interval]	
_Iagegroup_1	.4103218	.1286409	-2.84	0.004	.2219538	.7585541
_Iagegroup_2	.5555156	.1742744	-1.87	0.061	.3003725	1.027383
_cons	3.173913	.7589362	4.83	0.000	1.986362	5.071444

survive	Odds Ratio	Std. Err.	z	P>|z|	[95% Conf. Interval]	
_Icomorbids_2	.095546	.0419455	-5.35	0.000	.0404136	.2258902
_Icomorbids_3	.0853859	.0376076	-5.59	0.000	.0360147	.2024381
_cons	12.42857	4.882883	6.41	0.000	5.754461	26.8434

survive	Odds Ratio	Std. Err.	z	P>|z|	[95% Conf. Interval]	
qol	1.014869	.0049544	3.02	0.002	1.005205	1.024627
_cons	.7819161	.2427197	-0.79	0.428	.4255328	1.43677

Univariate analysis

```
. xi:stepwise, pr(.2) pe(0.05):logistic survive i.comorbids qol i.agegroup
i.comorbids     _Icomorbids_1-3   (naturally coded; _Icomorbids_1 omitted)
i.agegroup      _Iagegroup_0-2    (naturally coded; _Iagegroup_0 omitted)
                begin with full model
p = 0.9167 >= 0.2000  removing qol

Logistic regression                          Number of obs   =        300
                                             LR chi2(4)      =      55.91
                                             Prob > chi2     =     0.0000
Log likelihood = -165.65322                  Pseudo R2       =     0.1444
```

Multi-variate analysis

survive	Odds Ratio	Std. Err.	z	P>|z|	[95% Conf. Interval]	
_Icomorbids_2	.104202	.0461686	-5.10	0.000	.0437256	.2483225
_Icomorbids_3	.0906942	.0401646	-5.42	0.000	.0380731	.2160432
_Iagegroup_2	.6181633	.2101038	-1.42	0.157	.3175363	1.203408
_Iagegroup_1	.5630112	.19017	-1.70	0.089	.290405	1.091516
_cons	17.07745	7.565029	6.41	0.000	7.167284	40.69034

```
. estat gof

Logistic model for survive, goodness-of-fit test

          number of observations =        300
    number of covariate patterns =          9
            Pearson chi2(4) =             2.61
               Prob > chi2 =           0.6248
```

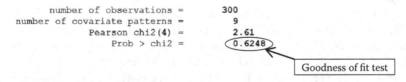

Goodness of fit test

FIGURE 5.20
Stata output for a binary logistic regression model.

variables. For the multi-variate analysis, we type 'xi:stepwise, pr(.2) pe(0.05):logistic survive i.comorbids qol i.agegroup'. As we can see from Figure 5.20, Qol is no longer significant in the final model, indicating that it is possibly a confounding variable. Unlike the linear regression model, the logistic regression model does not require many assumptions to be fulfilled. The appropriateness of model fit can be assessed by typing 'estat gof'. From the resulting output shown, we can see that there is no significant evidence to reject the null hypothesis that the model fits the data well ($p = .625$).

5.3.3.4 Predicted Probability

Sometimes, researchers may be interested in calculating the predicted probability of the outcome after the final multi-variate model. This is often used in deriving prognostic indices such as the Acute Physiology and Chronic Health Evaluation 3 score, which is used to predict the hospital mortality risk for critically ill-hospitalised adults (Knaus et al., 1991), and the Framingham Heart study, from which risk calculators (e.g. Risk Assessment Tool for Estimating Your 10-year Risk of Having a Heart Attack, available at http://cvdrisk.nhlbi. nih.gov/. Date accessed: 8 December 2014) for 10-year mortality have been developed. In Stata, the predicted probability can easily be obtained after running the final logistic regression model by typing 'predict p'. A new column of predicted probabilities, p will then be created and appended at the end of the dataset. The relationship between p and the explanatory variables (i.e. age group and comorbids) is given through a logit function, as shown in the following equation:

$$\log\left(\frac{p_i}{1-p_i}\right) = 2.8 - 0.6\text{age1}_i - 0.5\text{age2}_i - 2.3\text{comorbid1}_i - 2.4\text{comorbid2}_i \quad (5.1)$$

where:
p_i is the predicted probability of 6-month survival for each subject i
age1 and age2 are indicator variables taking on the value of 1 when the subject falls within the age group values of '40–64 years' or '65 and above', respectively, and 0 otherwise
comorbid1 and comorbid2 are indicator variables taking on the value of 1 for comorbid value '2' or '3 and more', and 0 otherwise

The coefficients are obtained by taking the log-odds of the ORs in multi-variate output in Figure 5.20.

5.3.3.5 Ordinal Outcome Measure–Ordinal Logistic Regression

A gastroenterologist wanted to evaluate a new device that was used to perform upper gastrointestinal endoscopy among his/her patients, with patient satisfaction being the key outcome measure. The satisfaction scores ranged from 1 (completely unsatisfied) to 5 (completely satisfied), and the comparison group was those on standard treatment. He/she also wanted to adjust for known confounders such as age and smoking. Because the outcome measure is ordinal in nature, we choose the ordinal logistic regression for the analysis (Table 5.1). In Stata, we type 'xi:ologit Satisfaction i.group age i.smoking, or'. Note that we have not used any stepwise method of 'automatically' selecting variables to be included in the multi-variate analysis, but rather 'forced' the variables age and smoking in the final model. This is sometimes done when clinicians feel that there is overwhelming clinical evidence to adjust for these known confounders in the analysis. As we can see from Figure 5.21, there is no significant difference in the satisfaction

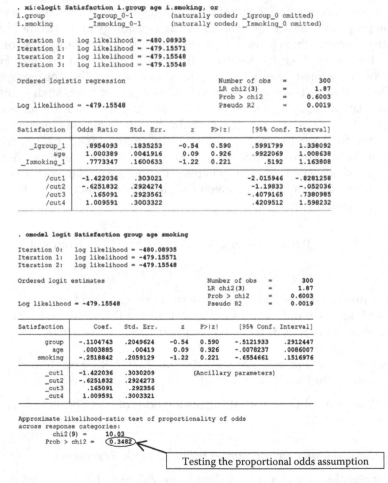

FIGURE 5.21
Stata output for an ordered logistic regression model.

between those on new treatment (group = 1) versus standard treatment (group = 0). The interpretation of the OR is a bit more convoluted: for those on the new device, the odds of being completely satisfied versus the other categories of satisfaction is 0.9 times lower, given the other factors such as age and smoking status are kept constant.

The main assumption in an ordinal logistic regression is proportional odds, that is, the coefficients that describe the relationship between levels 5 and 4 in the satisfaction scale is the same as that which explains the relationship between levels 4 and 3, and so on. To test the proportional odds assumption, one can download a user-written Stata program (type 'findit omodel') and then type 'omodel logit Satisfaction group age smoking'. As we can see from Figure 5.21, the proportional odds assumption is not violated ($p = .348$) in our example.

5.3.3.6 Categorical Outcome Measure–Multinomial Logistic Regression

The chief operating officer in a tertiary hospital in Melbourne wanted to explore whether age and presence of comorbid conditions had an impact on the place of discharge among patients hospitalised for acute stroke. Place of discharge was coded as 1 – home, 2 – rehabilitation centre and 3 – death. As there are three distinct levels for the categorical outcome variable, from Table 5.1 we see that a multinomial logistic regression is appropriate.

In Stata, type 'xi:mlogit Place age, base(1) rrr'. From Figure 5.22, we see that the RRR of being discharged to a rehabilitation centre increases by a factor of 1.23 for every year increase in age. Similar, the RRR for death is 1.28 for each year increase in age compared to being discharged to home. If one wanted to compare death with discharge to rehabilitation centre, then the base can be changed to (2). For number of comorbidities, the results are mixed, with three or more comorbidities being significant for rehabilitation centre versus hospital, and two comorbidities significant for death (Figure 5.22).

5.3.3.7 Count Data–Poisson Regression Model

A physiotherapist wanted to evaluate the effectiveness of a new program he/she implemented in a community hospital among patients who had a high risk of falling. The main outcome was the number of falls over a 6-month period. He/she wanted to make a comparison with an equivalent hospital that did not have the program, with adjustments made for confounders such as age and gender as well as taking into account the length of period that patients have been hospitalised for. From Table 5.1, it was determined that a Poisson regression model would be appropriate. In Stata, we type 'xi: poisson falls i.hospital age i.gender, exposure(patient_days) irr'. After adjusting for age and gender, the IRR of falls for the control hospital (coded 1) was 1.64 (95% CI: 1.48–1.82) compared to the intervention hospital (coded 0), and the result was statistically significant

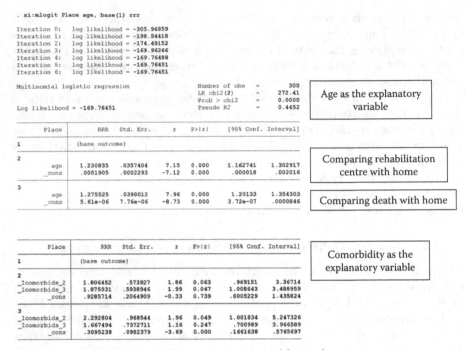

FIGURE 5.22
Stata output for a multinomial logistic regression model.

($p < .001$) (Figure 5.23). A goodness of fit test can be obtained by typing 'estat gof'. As we can see, the data does not fit the model well, $p < .001$, possibly due to over-dispersion. In this case, a negative binomial model can be fit by typing 'xi: nbreg falls i.hospital age i.gender, exposure(patient_days) irr'. There is little change to the effect size or p-value, although the standard error has increased.

5.3.3.8 Survival Data–Cox Regression Model

A cardiologist wanted to evaluate whether the presence of CHD was associated with survival among his/her cohort of patients. Survival was defined as the time from admission to hospital till death or the last known follow-up date, with the main outcome thus being death. In addition, he/she wanted to make adjustments for the following potential risk factors: sex, age and presence of diabetes. From Table 5.1, we know that a Cox regression model is needed. In Stata, firstly we need to declare the survival time by typing 'stset followupdays,

```
. xi: poisson falls i.hospital age i.gender, exposure( patient_days) irr
i.hospital          _Ihospital_1-2      (naturally coded; _Ihospital_1 omitted)
i.gender            _Igender_1-2        (naturally coded; _Igender_1 omitted)

Iteration 0:   log likelihood = -1179.3145
Iteration 1:   log likelihood = -1179.3145

Poisson regression                          Number of obs   =        300
                                            LR chi2(3)      =     342.43
                                            Prob > chi2     =     0.0000
Log likelihood = -1179.3145                 Pseudo R2       =     0.1268
```

Poisson model

```
       falls |      IRR    Std. Err.      z    P>|z|     [95% Conf. Interval]
-------------+----------------------------------------------------------------
_Ihospital_2 |  1.641089    .0856169    9.49   0.000     1.481577    1.817775
         age |  1.000671    .0010689    0.63   0.530     .9985784    1.002769
  _Igender_2 |  1.075176    .0302275    2.58   0.010     1.017534    1.136084
       _cons |  .1189033    .005379   -47.07   0.000     .1088145    .1299276
ln(patien~s) |        1   (exposure)
```

```
. estat gof

        Deviance goodness-of-fit  =    979.8907
        Prob > chi2(296)          =      0.0000

        Pearson goodness-of-fit   =   1083.604
        Prob > chi2(296)          =      0.0000
```

Goodness of fit test

```
Negative binomial regression                Number of obs   =        300
                                            LR chi2(3)      =      84.01
Dispersion    = mean                        Prob > chi2     =     0.0000
Log likelihood = -1028.7824                 Pseudo R2       =     0.0392

       falls |      IRR    Std. Err.      z    P>|z|     [95% Conf. Interval]
-------------+----------------------------------------------------------------
_Ihospital_2 |  1.634282    .1526655    5.26   0.000     1.360857    1.962644
         age |    1.0008    .0019193    0.42   0.677     .9970452    1.004569
  _Igender_2 |  1.097896    .0583563    1.76   0.079     .9892758    1.218442
       _cons |  .1266478    .0104013  -25.16   0.000     .1078178    .1487664
ln(patien~s) |        1   (exposure)
-------------+----------------------------------------------------------------
     /lnalpha |  -1.950889    .1189436                  -2.184014   -1.717764
-------------+----------------------------------------------------------------
       alpha |  .1421476    .0169075                    .1125887    .179467
------------------------------------------------------------------------------
Likelihood-ratio test of alpha=0:  chibar2(01) =   301.06 Prob>=chibar2 = 0.000
```

Negative binomial model

Testing for over-dispersion

FIGURE 5.23
Stata output for a Poisson regression model.

failure(died)'. Then, for the univariate Cox regression model, we type 'xi: stcox age'. As we can see from Figure 5.24, the hazard ratio (HR) for age is 1.0009, indicating that the HR of death increased by a factor of 1.0009 for every year increase in age. However, this was not statistically significant ($p = .762$). Similarly, we run the univariate analysis with the rest of the explanatory variables. Only presence of CHD and diabetes was significant. Those with CHD were 6 times more likely to die compared to those without CHD.

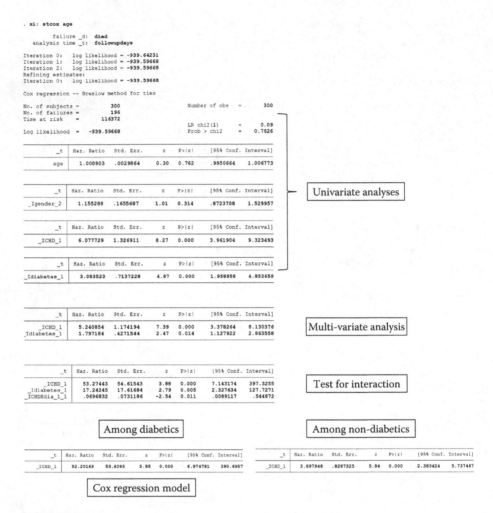

FIGURE 5.24
Stata output for a Cox regression model.

For the multi-variate analysis, we specify the following model: 'xi: stcox
i.CHD i.diabetes', because only one other variable other than CHD was
significant in the univariate analysis, and we do not really need to specify any
stepwise model selection criteria. As we can see from Figure 5.24, both CHD
and diabetes remained significant, indicating that both of these variables were
independently associated with mortality. At this stage, it may be worthwhile
to examine for interaction (or moderation) effects, that is, to explore whether
the relationship between CHD and mortality is different for those with and
without diabetes. To implement this in Stata, we type 'xi: stcox i.CHD*i.
diabetes'. Figure 5.24 shows that the result of the test for interaction was
significant ($p = .011$). As a consequence, one should present the results for the

relationship between CHD and mortality separately for diabetics and non-diabetics. This can be easily achieved in Stata by typing 'xi: stcox i.CHD if diabetes==1' and 'xi: stcox i.CHD if diabetes==0', respectively. The results show that there was a significantly higher HR for CHD among diabetics (HR = 52) versus those without diabetes (HR = 4).

5.3.3.9 *Generalised Estimating Equations*

In Stata, there is a rich suite of commands under the generalised estimating equation (GEE) modelling framework that allows for the study of outcomes or explanatory variables, which may be repeated over time. The outcome variables can be continuous, binary or count data. GEE models are often seen as extensions of the usual generalised linear models but are flexible enough to handle non-independence in the data (Hardin & Hilbe, 2013). In particular, the models are useful for clustered or longitudinal data. When running the GEE model, three main things need to be specified in Stata:

1. Link function (e.g. identity for continuous outcomes, log for count data and logit for binary data)
2. Distribution family (e.g. Gaussian, Binomial, Poisson, etc.)
3. Correlation matrix (e.g. exchangeable, unstructured, fixed)

In Stata, we type 'xtgee depvar indepvars, family() link() corr()'. We can add the xi prefix to indicate the presence of categorical independent variables. We need to use the xtset command prior to xtgee to set up data in a panel format, for example xtset panelvar timevar. Alternatively, we can specify 'xtgee depvar indepvars, family() link() corr() i(panelvar) t(timevar)' each time we run a model, but this is obviously more cumbersome.

If the model is correct and the correlation matrix mis-specified, the parameter estimates remain consistent. However, modelling the correlation may boost efficiency. The correlation matrix can be specified in the following ways:

1. *Independence.* Correlation between observations = 1
2. *Exchangeable.* Compound symmetry
3. *Unstructured.* Every pair can have a different value
4. *Autoregressive.* Correlation decays as observations are further apart
5. *Stationary or non-stationary.* Correlation matrix as a function of time
6. *Fixed.* User-specified matrix

For panel data with no inherent ordering in observations, we may use the exchangeable correlation matrix. For a small number of repeated measures, we can use the unstructured matrix, and for observations with a temporal element (e.g. disease surveillance data), one may use the autoregressive or stationary/non-stationary matrices.

5.3.3.10 Clinical Examples

A gastroenterologist tested a cohort of 100 patients annually for a period of 6 years in the presence of *Helicobacter pylori* bacteria and also measured their QoL. He/she was interested in testing the two main hypotheses: Firstly, QoL was not associated with age after adjusting for gender of patient. The second hypothesis was that there were significant changes in the proportion of infections over the years after accounting for sex and age. We need to format the dataset from a wide to a long format (Figure 5.25) by typing 'reshape long qol infection, i(subjectid) j(visit)', and then set the panel data information with 'xtset subjectid visit', before specifying the GEE model.

Because QoL is continuous in nature and we can expect a temporal correlation in QoL, we specify a Gaussian distribution with an identity link and AR1 as the variance–covariance matrix. In Stata, we type 'xi:xtgee qol age i.sex, family(gaussian) link(identity) corr(ar1)'. As we can see from Figure 5.25, age was negatively associated with QoL (coefficient = −0.02), but this was not statistically significant, after adjusting for sex ($p = .787$). Notice how the standard errors are larger than a naïve model that assumes no correlation within subjects (i.e. independent or identity variance–covariance matrix).

The second hypothesis involves a binary outcome measure (infection: yes/no), and hence, we specify a binomial distribution under family and a logit link function. If we do not know any *a priori* structure in the correlation, we can specify the 'exchangeable' correlation matrix. In Stata, we type 'xi:xtgee infection i.visit i.sex age, family(binomial) link(logit) corr(exch) eform'. The 'eform' option allows us to look at the OR instead of the coefficients. As we can see from Figure 5.26, the OR of *H. pylori* infection is significant at years 5 and 6, with an OR of 2.8 and 2.9, respectively, compared to the baseline year ($p < .001$).

There is a distinction between GEE and mixed models (also known as random effects models), with the former looking at population averages, and the latter being subject specific. There is currently no co-census on which modelling approach is more appropriate, although the GEE has been reported to be more appropriate for estimating the associations between neighbourhood risk factors and health (Hubbard et al., 2010). The mixed modelling approach is not discussed here, but interested readers can consult the textbook *Multilevel and Longitudinal Modeling Using Stata*, which provides the methodological background of the various models, including the use of Stata software to address practical research problems (Rabe-Hesketh & Skrondal, 2008). Accompanying datasets are also provided in the website http://www.stata.com/bookstore/multilevel-longitudinal-modeling-stata/ (Date accessed: 3 December 2015).

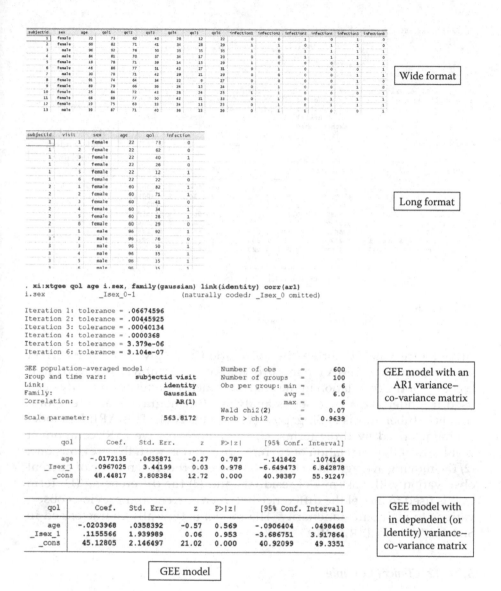

```
. xi:xtgee qol age i.sex, family(gaussian) link(identity) corr(ar1)
i.sex           _Isex_0-1        (naturally coded; _Isex_0 omitted)

Iteration 1: tolerance = .06674596
Iteration 2: tolerance = .00445925
Iteration 3: tolerance = .00040134
Iteration 4: tolerance = .0000368
Iteration 5: tolerance = 3.379e-06
Iteration 6: tolerance = 3.104e-07

GEE population-averaged model          Number of obs      =       600
Group and time vars:    subjectid visit     Number of groups   =       100
Link:                       identity        Obs per group: min =         6
Family:                     Gaussian                       avg =       6.0
Correlation:                AR(1)                           max =         6
                                               Wald chi2(2)     =      0.07
Scale parameter:            563.8172           Prob > chi2      =    0.9639
```

qol	Coef.	Std. Err.	z	P>|z|	[95% Conf. Interval]
age	-.0172135	.0635871	-0.27	0.787	-.141842 .1074149
_Isex_1	.0967025	3.44199	0.03	0.978	-6.649473 6.842878
_cons	48.44817	3.808384	12.72	0.000	40.98387 55.91247

qol	Coef.	Std. Err.	z	P>|z|	[95% Conf. Interval]
age	-.0203968	.0358392	-0.57	0.569	-.0906404 .0498468
_Isex_1	.1155566	1.939989	0.06	0.953	-3.686751 3.917864
_cons	45.12805	2.146497	21.02	0.000	40.92099 49.3351

GEE model

FIGURE 5.25
Stata output from a GEE analysis (continuous outcome).

5.3.3.11 ARIMA Models

Sometimes, count data of health outcomes or indices are presented serially over time, and some form of temporal correlation and seasonality may be inherent in the data (e.g. routinely collected infectious disease

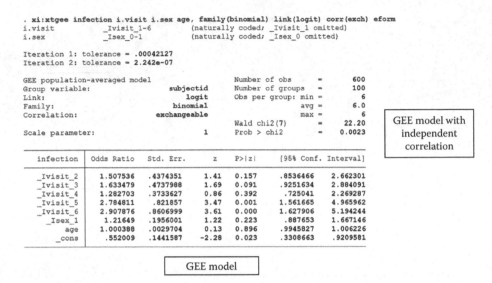

```
. xi:xtgee infection i.visit i.sex age, family(binomial) link(logit) corr(exch) eform
i.visit              _Ivisit_1-6       (naturally coded; _Ivisit_1 omitted)
i.sex                _Isex_0-1         (naturally coded; _Isex_0 omitted)

Iteration 1: tolerance = .00042127
Iteration 2: tolerance = 2.242e-07

GEE population-averaged model            Number of obs     =        600
Group variable:               subjectid  Number of groups  =        100
Link:                             logit   Obs per group: min =         6
Family:                        binomial                  avg =       6.0
Correlation:               exchangeable                  max =         6
                                         Wald chi2(7)      =      22.20
Scale parameter:                       1 Prob > chi2       =     0.0023
```

GEE model with independent correlation

infection	Odds Ratio	Std. Err.	z	P>|z|	[95% Conf. Interval]	
_Ivisit_2	1.507536	.4374351	1.41	0.157	.8536466	2.662301
_Ivisit_3	1.633479	.4737988	1.69	0.091	.9251634	2.884091
_Ivisit_4	1.282703	.3733627	0.86	0.392	.725041	2.269287
_Ivisit_5	2.784811	.821857	3.47	0.001	1.561665	4.965962
_Ivisit_6	2.907876	.8606999	3.61	0.000	1.627906	5.194244
_Isex_1	1.21649	.1956001	1.22	0.223	.887653	1.667146
age	1.000388	.0029704	0.13	0.896	.9945827	1.006226
_cons	.552009	.1441587	-2.28	0.023	.3308663	.9209581

GEE model

FIGURE 5.26
Stata output from a GEE analysis (binary outcome).

surveillance data). Ignoring the correlation may give rise to incorrect standard errors. ARIMA models provide a convenient way to specify and account for the correlation in the analysis. Such a model has recently been used to undertake a time series analysis of demographic and temporal trends of tuberculosis in Singapore (Wah et al., 2014). This ARIMA model is characterised by three terms: (1) the autoregressive (AR) component that looks at the correlation of current observation with the previous one, (2) the moving average (MA) term that looks at the dependence of current observation with past model residuals and (3) the differencing (D) term that is used to model the difference between current and previous observations to maintain stationarity. The model is typically characterised in Stata as 'arima (AR,D,MA)'.

5.3.3.12 Clinical Example

A public health physician wanted to examine the relationship between monthly dengue fever (DF) notifications and the Ovitrap index (based on counts of *Aedes albopictus* mosquitoes trapped) in a small community in Brisbane, Australia. Serial monthly data were available to him/her for a period of 3 years. It was decided that the ARIMA model would be the most appropriate method. Firstly, we need to set up the data in a time series format by typing 'tsset month', which indicates to Stata that month is the variable capturing time information. Next, it is a good idea to visually examine the data series, and also look at the autocorrelation and partial autocorrelation (PAC) plots, which give an indication on how to specify the number of parameters for the MA and AR terms in the ARIMA model. The corresponding commands are `scatter dengue month, c(l), ac dengue` and `pac dengue`, respectively.

As we can see from Figure 5.27, DF notifications appear to be correlated across the months, and the PAC plot suggests an AR value of 1 (based on lag values extending beyond the 95% CI band in shaded area) and a MA value of 2. The series appears stationary, so we do not need to take the difference between successive observations and model the difference instead. Formal tests for stationarity exist in Stata (type 'help dfuller'). To run a basic ARIMA model with AR = 1 and MA = 2, we type 'arima dengue ovitrap, arima(1,0,2)'. The results indicate that there is a positive and significant association between DF notifications and Ovitrap index (coefficient = 0.74, $p < .001$) after accounting for serial correlation in the data.

5.4 Issues to Note

5.4.1 Crossover Trials

The design of a crossover trial was discussed in Chapter 2. The analytical method is discussed here. Suppose we have a two-treatment (Active and Placebo) and two-period (Time 1 and Time 2) study as shown in Table 5.3. Assume the outcome is pain score measured on a continuous scale. The mean difference in groups A and B are $\bar{y}$ and $\bar{z}$, respectively. To test for treatment effect (i.e. change in pain score between the Active and Placebo groups), we compare the difference between $\bar{y}$ and $\bar{z}$ using an independent Student's t-test, assuming there is no treatment-by-period interaction effect. To test for period effect (i.e. difference in treatment effect across Time periods 1 and 2), we test for $(\bar{y} + \bar{z})$. A test for interaction can also be performed, and where significant, data from the first time point, Time 1 can be used in the analysis. Extensions to more than two groups, treatments and sequences as well as other response variables besides continuous ones are discussed in Armitage's book (Armitage et al., 2002).

5.4.2 Cluster RCT

In cluster RCTs, the response of subjects within each cluster is expected to be correlated compared to subjects across clusters. The analytical method needs to account for this clustering. GEE models, discussed in Section 5.3.3.9, are able to account for the various forms of correlations within the clusters, although other multi-level models such as generalised linear latent and mixed models are also available (Rabe-Hesketh & Skrondal, 2008).

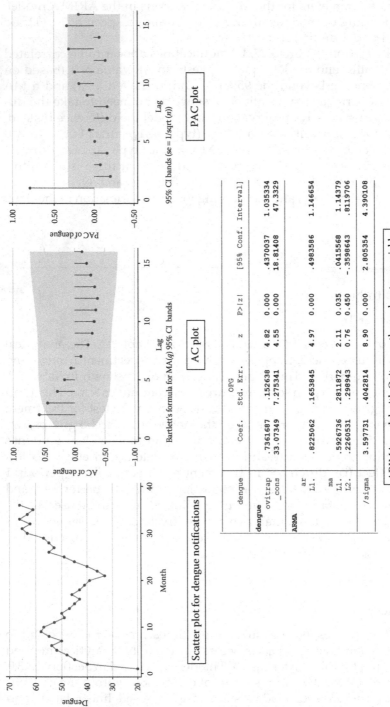

FIGURE 5.27
Stata output from an ARIMA analysis.

TABLE 5.3

Design and Analysis of a Crossover Trial

Group	Time 1		Time 2	Mean Difference
A	Active $\overline{x}_1$	Wash-out	Placebo $\overline{x}_2$	$y = \overline{x}_1 - \overline{x}_2$
B	Placebo $\overline{x}_3$		Active $\overline{x}_4$	$z = \overline{x}_3 - \overline{x}_4$

5.4.3 Intention-to-Treat versus Per-Protocol Analysis

Sometimes, in a trial, subjects drop out of the study due to various reasons such as death and refusal to continue to participate due to ill-health. Some may not follow the trial protocol strictly (i.e. not taking medication as prescribed, stop taking medication or even switch over to the other arm of the study). One option is to leave this group out in the analysis (i.e. only analyse the subset who provided complete data on the key outcome measures). This is known as the per-protocol approach, which has a major limitation; specifically when those excluded are systematically different from those who completed the trial. This type of analysis evaluates what happens when patients actually take the drug or intervention as opposed to being offered the drug. Alternatively, the procedure of including all patients in the groups that they were originally randomised to is known as the intention-to-treat (ITT) method. This method is more applicable at the clinical setting when clinicians are deciding to choose between treatments. It is usually used to study effectiveness of an intervention (regardless of patient's adherence to allocated treatment arm), while efficacy is established in the per-protocol analysis, usually with the assumption of an ideal situation where patients are compliant to treatment and protocol. The ITT procedure reduces the bias associated with non-random dropouts.

5.4.4 Missing Data

Missing data are common in clinical research, especially in observational studies where data have been collected retrospectively. For instance, information may have been collected from patients' case notes at hospitals, and not all physicians may have ordered the tests or diligently recorded them in the case notes or database. The amount of missing data and the nature of missingness will have an impact on the analysis, results and interpretation of the study. Missing completely at random is when the missingness is not related to the treatment or outcome measures. For example, patients not attending the hospital because of reasons not relating to their health. In this scenario, it is fine to analyse all available samples.

Missing not at random occurs when there is a systematic reason for the missingness, but this is not captured in the existing data and cannot be explained. In such a situation, it is difficult to analyse and interpret the study other than to find out reasons why the responses were missing (i.e. through interviews and follow-ups). Missing at random occurs when the probability of withdrawal/ dropout is related to the intervention, baseline or follow-up outcome measurements, for example patients may drop out when they experience adverse effects of a new drug being studied and so on. There are several missing data imputation techniques to address this situation, and these are discussed extensively in the following website: http://missingdata.lshtm.ac.uk/ (Data accessed: 12 December 2014), providing discussions on nature of missingness, statistical approaches, software and examples. Other methods also exist, including tensor factorization-based methods for QoL questionnaires (Garg et al., 2014). It should be noted that analysis of clinical disease registries poses a number of problems, particularly in terms of missing data and inadequate risk adjustment when comparing outcomes across groups of patients.

5.4.5 Multiplicity

Multiplicity occurs in research when there are a large number of end points being analysed, large number of subgroups analysed, multiple times the dataset is analysed or analysis performed with a large number of covariates. For example, in a psychiatric RCT, besides various psychiatric assessment tools such as the Beck Depression Inventory, Brief Psychiatric Rating Scale and Mini-Mental State Examination, length of stay and QoL outcome measures may have been collected and analysed. At the typical level of significance (or type 1 error or α) of 5%, we would expect five out of every hundred tests to give rise to a significant result by chance alone. The simplest method to rectify this in the analysis stage is to use a Bonferroni correction method. We calculate a new $\alpha_i = \alpha/n$, where n is the number of comparisons to be made or tests to be performed. For instance, if there are 10 subgroup analyses to be performed, the new level of significance against which the p-value from the experiment should be evaluated is a more stringent $\alpha = 0.05/10 = 0.005$. For large number of comparisons, especially in gene expression datasets, it is more feasible to use alternate methods such as the false discovery rate, which is described here (Motulsky, 2010). Multiple testing is also present when baseline differences in demographic and other clinical variables are tested between the experimental and control groups.

5.5 Key Questions to Ask a Clinician

1. What is the main aim and related hypothesis?
   A statistical test should be written for each aim/hypothesis.
2. Is the study adequately powered and do you have issues with recruitment/ subject accrual/retention?
   Non-parametric methods should be considered for small samples and subgroup analyses avoided where possible.

3. What is the main outcome measure, and what is the scale of measurement?

   This will have an impact on the choice of the analysis method.

4. What is the nature of the other variables measured in the hypothesis being tested?

   This will affect the statistical model to be used.

5. How many explanatory variables are there in the study?

   The use of multi-variate models and adjustment for multiplicity need to be considered.

6. Is there any bias or measurement error in the variables?

   Alternative sources of information/variables should be considered in the presence of excessive bias and models that incorporate measurement error should be evaluated.

7. Are there subgroups that you want to compare?

   Consider power issues as well as multiplicity.

8. How many hypotheses do you have in the study?

   Consider issues of multiplicity and data dredging.

9. Do you have multiple endpoints in the study?

   Consider issues of multiplicity and data dredging.

10. What are the possible confounding variables in the study and have they been collected in the study?

    Multi-variate models should be considered in the analysis.

11. Are there missing data in the dataset, and what is the nature of missingness, if any (i.e. are they systematic)?

    Consider missing data imputation techniques.

12. Are the observations independent? If not, in what ways are they clustered or repeated?

    Consider GEE or mixed models.

13. Is the database collected in a single spreadsheet, in a case by variable format?

    Request for data to be provided in the preferred format and avoid undertaking excessive data management tasks.

5.6 A Collaborative Case Study

Project title: Disability impacts length of stay in general internal medicine patients.

Collaborating clinician: Dr. Ng Yee Sien, MBBS, MRCP (London), FAMS; head and senior consultant, department of rehabilitation medicine, Singapore General Hospital, Singapore.

Output: Following publication. Disability impacts length of stay in general internal medicine patients. (Tan et al., 2014).

Study summary: The study's main aim was to investigate the relationship of disability with LOS among internal medicine patients after controlling for comorbidity. Disability was measured using the Functional Independence Measure (FIM) recorded at discharge. Comorbidity was measured using the Charlson Comorbidity Index. The study found in the multi-variate analysis that variables independently associated with longer LOS included the motor FIM score ($p < .001$), presence of social issues such as caregiver unavailability ($p < .001$), non-realistic patient expectations ($p = .001$) and administrative issues impeding discharge ($p = .016$).

Analysis plan: Because LOS was measured on a continuous scale, and we were interested in more than one independent variable, we can see from Table 5.1 that the linear regression model is indeed most appropriate. On closer examination of LOS, we found that it was positively skewed, and hence, a natural logarithmic transformation was performed, as reported in the manuscript. In preliminary discussions, it was suggested that we dichotomise LOS into two groups (outliers vs. non-outliers based on a threshold value), but it was decided that there was less clinical relevance to do so, and this was not pursued. In terms of the race variable, we grouped them into Chinese versus Others (Indian, Malays and Eurasians). Looking back, it may have been better to leave them in their original categories to see whether there were finer differences between the minority races in terms of relationship between disability and LOS.

Key Learning Points

1. Learn how to choose the most appropriate statistical technique for the analysis.

2. Acquire skills on formulating the specific hypothesis in discussion with the collaborator, running the test in Stata and interpreting the output.

3. Understand how the SAP is written in grants and manuscripts.

4. Learn about the common issues involved in the analysis of clinical research datasets, such as missing data and multiplicity.

5. Use a checklist of 'questions to ask the collaborator', so that decisions on the optimal statistical test to employ can be made relatively quickly.

References

Armitage, P., Berry, G., & Matthews, J. (2002). *Statistical Methods in Medical Research* (4th ed.). Oxford: Blackwell Science.

Garg, L., Dauwels, J., Earnest, A., & Leong, K. P. (2014). Tensor-based methods for handling missing data in quality-of-life questionnaires. *IEEE J Biomed Health Inform*, *18*, 1571–1580.

Hardin, J., & Hilbe, J. (2013). *Generalized Estimating Equations* (2nd ed.). Boca Raton, FL: Chapman & Hall/CRC Press.

Hubbard, A. E., Ahern, J., Fleischer, N. L., Van der Laan, M., Lippman, S. A., Jewell, N., ... & Satariano, W. A. (2010). To GEE or not to GEE: Comparing population average and mixed models for estimating the associations between neighborhood risk factors and health. *Epidemiology*, *21*, 467–474.

Knaus, W. A., Wagner, D. P., Draper, E. A., Zimmerman, J. E., Bergner, M., Bastos, P. G., & Damiano, A. (1991). The APACHE III prognostic system: Risk prediction of hospital mortality for critically ill hospitalized adults. *Chest*, *100*, 1619–1636.

Landis, J. R., & Koch, G. G. (1977). The measurement of observer agreement for categorical data. *Biometrics*, *33*, 159–174.

Motulsky, H. (2010). *Intuitive Biostatistics: A Nonmathematical Guide to Statistical Thinking* (2nd ed.). New York: Oxford University Press.

Rabe-Hesketh, S., & Skrondal, A. (2008). *Multilevel and Longitudinal Modeling Using Stata*. College Station, TX: Stata Press.

Tan, C., Ng, Y.S., Koh, G.C., De Silva, D.A., Earnest, A., Barbier, S. (2014). Disability impacts length of stay in general internal medicine patients. *J Gen Intern Med.* *29*(6), 885–90. doi:10.1007/s11606-014-2815-z.

Wah, W., Das, S., Earnest, A., Lim, L. K., Chee, C. B., Cook, A. R., ... & Hsu, L. Y. (2014). Time series analysis of demographic and temporal trends of tuberculosis in Singapore. *BMC Public Health*, *14*, 1121.

6

Effective Communication Skills

6.1 Introduction

Improved communication between biostatisticians and clinicians will help foster more productive collaborations, and this would help to make the process more smooth and enjoyable for all parties involved. In our survey among biostatisticians (Chapter 11), it was found that 78% of biostatisticians identified 'communication between biostatisticians and clinicians' as an issue to address in order to enhance greater collaborations. In this chapter, we provide practical suggestions and clear advice on how to enhance communication skills for biostatisticians, and this includes the specification of timelines, milestones and clarification of the allocation of tasks in a collaborative project. We start with the initial meeting with the collaborator, where important issues such as the scope of collaboration are discussed and established during the conversation. The use of a customised information collection sheet is recommended to document and help manage the expectations that can arise during the communication process and also to increase efficiency. Other unique contributions in this chapter include tips on presenting statistics effectively to a non-statistical audience, for example, through drawings, Venn diagrams and so on, as well as addressing difficulties in understanding medical jargon and acquiring good habits in statistical presentation (oral and poster). Towards the end of the chapter, links to useful online resources as well as an interview with an experienced researcher on improving the communication process with biostatisticians are presented.

6.2 The Initial Meeting

It is important that the biostatistician takes steps to prepare for the first meeting with the collaborator to get the most out of the meeting and also to make an impact. As the saying goes, first impression lasts. If possible, a list of information can be requested from the collaborator prior to the meeting.

This would help to make the consultation efficient and also help to manage any expectation on outputs arising from the meeting. An example of such a checklist is displayed in Figure 6.1. The first thing to establish is the scope of collaboration. It is easier to communicate and talk about the project when the scope is made clear. Grant proposals, manuscripts or statistical consultations may require different types of inputs and deliverables from the biostatistician. For instance, if the request is for a biostatistical input in a grant proposal, it may be necessary to perform sample size calculation and write the statistical analysis plan, whereas for a manuscript submission, the discussion may primarily revolve around hypothesis testing and statistical methods of analysis.

Other types of requests such as institutional review board (ethics) applications, review of statistical methods applied in a journal article or even production of tables and graphs for conference proceedings may be consolidated under 'statistical consultation', where a time cap can be used to limit the amount of time the statistician spends on such projects. This is especially useful because such activities may not directly contribute towards the usual key performance indicators (grants and publications) in academic

Please indicate the nature of biostatistics collaboration

    1. Grant submission ☐   2. Publication ☐   3. Brief consultation (max 1 hour) ☐

For grants or publications, when is your intended submission date? _____

Indicate the specific input required (check all that apply):

    Data analysis   ☐

    Sample size calculation   ☐

    Manuscript writing   ☐

    Advice on study design, analysis plan, etc.   ☐

Please provide the following information prior to the meeting:

    1. Summary (max 2 pages) of the title, study aims and hypothesis, proposed study design and analytical methods (if any)

    2. For sample size calculations, descriptive statistics like mean, standard deviation, frequency counts and clinically meaningful effect size that needs to be studied.

    3. For grant submission or publication, a copy of the latest draft document

    4. Any other relevant literature or information, highlighting the key aspects that need discussion or clarification

FIGURE 6.1
Checklist of collaboration request.

institutions. Identifying the specific input required (e.g. statistical analysis or sample size calculation) also ensures that the initial meeting ends up productive with some clear deliverables outlined and possibly delivered after the initial meeting or within a suitable period of time thereafter. It also provides an opportunity for the biostatistician to read up beforehand and prepare the relevant materials for discussion during the meeting. This is particularly useful if the analytical approach is going to be complex (e.g. the collaborator may indicate that he/she would like to apply some form of Bayesian adaptive study design to the proposed study). Any additional documents, such as a draft grant proposal, can also be sent to the biostatistician to read and better comprehend the clinical topic *a priori* to the meeting. For instance, an intensive care unit (ICU) specialist may be interested to develop a new method to risk stratify patients who end up in the hospital's ICU. With the information provided in the draft document, the statistician can do some simple online searches and discover more about how current prognostic algorithms such as the Acute Physiology and Chronic Health Evaluation II scores have been developed and, more importantly, identify and get accustomed to the statistical models behind the algorithms (e.g. the logistic regression model in this example). With the additional information, the statistician would be able to better relate to the clinician's discussion and hence confidently communicate with the clinician on the appropriate choice of statistical model to be used in the project.

Detailed information on aims and specific hypotheses is critical for the biostatistician to provide sound advice on the appropriate statistical methods to apply or the correct sample size calculations to perform. Often, the initial draft of the hypothesis from the clinician may not be specific enough, and there needs to be several iterations before a workable hypothesis is written down. This negotiation process may take some time, and in order to minimise any possible frustration during this iterative process at the first consultation meeting, it is advisable that the statistician requests for this information prior to the first meeting. This will also allow the collaborator to have some time to think about the hypotheses and be adequately prepared for the initial meeting. For instance, a respiratory specialist may be interested to explore the effectiveness of a new drug X in terms of improving lung function among asthma patients. The initial hypothesis may be written along the lines of the following: 'Patients on drug X will show a significant improvement in their lung function capacity as compared to those on Salbutamol'. Although this is a good start, the hypothesis is not specific enough, and it lacks important details which would be needed to identify the appropriate statistical techniques to be used, or to calculate the sample size. In particular, critical information about the type of outcome variable (and measurement scale) as well as the clinically meaningful effect size to be tested is lacking. A much improved hypothesis would read as follows: 'Patients on drug X will show a significant 25% improvement in the mean peak expiratory flow from baseline to 6 months in their lung function capacity, as measured from a spirometer as compared to those on Salbutamol'. Another important point to clarify with the clinician in this example is whether the outcome is measured at a single time point or whether the aim is to examine the change in score over two time points. In our example, the latter would be true. This would have an impact on the way data are analysed and sample size calculations performed. In particular, the standard deviation of change scores is usually much smaller than that of scores measured at baseline or follow-up. Using standard deviation of outcome measured at baseline in calculating the sample size for change scores may give rise to a larger than the required sample size calculation in the planning of the study.

The starting point for discussion during the initial meeting can be the items provided in the checklist (Figure 6.1). Where possible, the statistician should ask the clinician to verbally provide a background to the research, objectives and hypotheses. There may be variations to the document initially provided, and this should be clarified. The statistician should listen intently to the collaborator and take down notes when necessary as some medical terminology may not be obvious initially, but can be easily looked up later. Although the statistician should not interrupt when the clinician when he/ she is describing the project, it is fine to stop the conversation and ask for clarification. This is especially when there is a medical terminology that is not easily understood, or if the clinician is veering off-topic. For instance, an enthusiastic collaborator from the infectious diseases speciality may refer to *Streptococcus pneumoniae* without describing what the condition actually is. Sometimes, statisticians may find it difficult to pronounce certain medical terminologies as they have not been regularly exposed to them, unlike the clinician. Rather than avoid using the word in the discussion, it may be better to use an abbreviated term instead. For instance, it is certainly easier to refer to 'juxtaglomerular complex' as JC during discussions with the collaborator, but this abbreviation needs to be communicated. Sometimes, the collaborator may veer off-topic, for instance talk about a future study on cost-effectiveness or even discuss related statistical techniques he/she has read from journals. It is important to lead the discussion back to the aims listed for the current project and to get this finalised before subsequent discussions on the appropriate statistical methods takes place. Otherwise, this would end up in a situation where one puts the cart in front of the horse, and this may lead to unnecessarily long (and sometimes frustrating) meetings.

Clinicians are busy people, who usually struggle to juggle their time between clinic and research, unless they have been allocated protected time to conduct research by their department or a funding agency. Inevitably, some may not be able to afford time to meet with the biostatistician for even an hour during office hours when they are supposed to run clinics. Some clinicians may feel that a 5–10 min consultation is sufficient for them, but in reality this is usually not enough. It usually takes at least 45 min to have a robust discussion on the research question, aims, hypothesis, methods,

tables and charts, timelines and even schedule follow-up meetings. If a collaborator requests for a brief meeting, it is better to reschedule that meeting
to another day when the collaborator is able to meet for a longer time that is
mutually convenient time for both the statistician and the clinician; sometimes, this may be arranged during lunchtimes or after work hours. In some
instances, the statistician may also need to visit the clinic to meet his/her
collaborator, which can be a positive development, as the statistician can
then get a better understanding of the nature of the illness and the treatment process by observing the care delivery process and the clinic environment. This advice was suggested by one of the clinicians interviewed in this
book (Chapter 11).

Some collaborators like to bring printed copies of related articles for discussion during the initial meeting. Their intention is usually well-meaning,
but they fail to realise that it takes time to read through the articles, even if
it was just related to (say) the methods section of the articles. These articles
(and any other relevant materials) should be sent to the statistician prior to
the meeting. If presented with a new article during the meeting, the biostatistician can request that the collaborator summarises the article and identifies the issues that need clarification from the article. A discussion of the
methods section of the manuscript can be performed after a brief period
of reading, but if more time is required, it is better to request this from the
collaborator, rather than rushing through the article. During the initial meeting, the statistician should try and obtain any additional information that
may be necessary for the timely completion of the project (e.g. effect size for
sample size calculations or a data dictionary or codelist for the variables in
the dataset that needs to be analysed). At the end of meeting, a proposed
action plan should be drawn up, which may include follow-up actions from
both parties, timelines, deliverables and scheduling of subsequent meetings
(if necessary).

If sufficient details and information have been obtained from the collaborator during the initial meeting, the statistician may propose a timeline for
the data analysis or input in grant application to revert to the collaborator.
This timeline needs to be realistic and mutually agreed upon. Collaborators
often fail to understand that the biostatistician usually works on a number
of other concurrent projects. If the analysis is fairly straightforward and not
too complex to explain, it can be sent to the collaborator via e-mail after it
has been completed. Grant proposals and draft manuscripts can similarly be
sent to collaborators, with input from the biostatistician marked under 'track
changes' or as comments. Otherwise, it may be prudent to set up another
meeting to explain the work, particularly if the statistical methods can be
challenging to explain. Alternatively, if the clinician is unable to meet up,
the statistical output can be sent via e-mail and followed up closely with a
telephone call to explain the analysis. It is important that the biostatistician
documents any data management or statistical processing request made by
the collaborator on the phone or meeting (e.g. recoding variables or dropping

observations) in writing (for instance as a comment in the statistical software command code), so as to ensure reproducibility and avoid misunderstandings in the future. It has also been said that 'a successful collaboration would depend on the intersection of the setting and the personalities involved' (personal communication: Professor Jay Herson, Johns Hopkins Bloomberg School of Public Health, Johns Hopkins University, Baltimore, Maryland). This would depend on the relative status of the clinician and statistician. If the clinician is, say, 60 years old, a full professor with international reputation and the biostatistician a 25-year-old master's degree just out of grad school, the situation will be different than if the backgrounds are reversed or identical and it is advisable for the biostatistician to take note of this.

6.3 Difficulty in Understanding Medical Jargon

Much medical jargon is notoriously difficult to pronounce and challenging to spell. It is a refreshing change for statisticians who are more used to statistical terms, symbols, formulae and equations that have been taught in school. Unless the statistician has undertaken a postgraduate course in biostatistics, epidemiology or public health, or worked in a hospital or health research environment, he/she may not be familiar with many of the medical terminologies. Usually, this difficulty is overcome through experience (i.e. understanding of medical jargon usually gets better with greater collaborative projects undertaken). In particular, statisticians who work in a specific medical field (e.g. oncology) may not have this problem after a period of time of being exposed to commonly used jargon (e.g. chemotherapy, cancer clinical staging, etc.) through interactions with oncologists and other health care professionals. For those who collaborate with medical professionals from a wider spectrum of specialities, understanding new medical jargon can be a real challenge. During conversations with clinicians on research projects, it is perfectly fine to ask for clarifications on medical terms and abbreviations (common ones include 'Dx' for diagnosis, 'Sx' for symptoms, 'Rx' for prescription and 'LOS' for patient's length of stay). Taking notes during conversation is helpful for remembering long names and those with complex spelling. Voice recorders can also be used to record conversations and the notes transcribed later on.

In addition, the communication process can be made more fruitful if the statistician obtains related documents prior to the meeting, and if some background research can be done on the topic. For instance, a quick search on the Internet about 'treatment for COPD' will result in information on tiotropium bromide, a drug used to treat patients with chronic obstructive pulmonary disease (COPD) and pulmonary rehabilitation, a multidisciplinary team intervention for COPD patients. This *a priori* knowledge will be useful in discussions with the respiratory clinician. Online tools as well as medical textbooks and journals can be consulted on the topic that will be the focus of discussion. For instance, WebMD is a useful website resource that aims to provide practical and relevant content source for health and medicine (http://www.webmd.com/. Date accessed: 27 January 2015). Some medical terms also have simpler layman explanations, and the collaborator should be encouraged to use these simpler terms. Examples include 'dry mouth' for xerostomia, and 'nose bleed' for epistaxis.

6.4 Challenges in Explaining Statistical Concepts to Clinicians

Just as statisticians sometimes find it difficult to understand medical terminology, clinicians similarly find some of the statistical terms and phrases used during discussions intimidating. It is the statistician's responsibility to explain the modelling strategy and any other related statistical terms to the collaborator in clear, simple enough language so that the collaborator understands and is able to explain it to his/her colleagues/others. Wherever possible, it would be advisable for the statistician to use simple language in describing statistical terms and phrases, avoid presenting equations and statistical symbols (unless they are absolutely necessary) and use figures or drawings to aid in the communication process. Some practical examples are provided in Sections 6.4.1 through 6.4.4.

6.4.1 Using the Venn Diagram

The Venn diagram is an elegant medium to visually describe multiple events or disease categories that overlap. For instance, a clinician may be interested in looking at the various types and categories of comorbidities as predictors of mortality. The Venn diagram allows one to visually look at the various categories, including intersections that may help in identifying the specific categories of variables to include in the model, deciding on the categories to collapse upon, and even in explaining the concept of confounding. One can easily see from the Venn diagram in Figure 6.2 that 7.7% of the patients do not have any of the comorbid conditions or conversely 92.3% have at least one. About 3.5% of the patients have all three conditions (i.e. hypertension, diabetes and heart disease). With such a small proportion of patients having all three conditions, one may then not decide to study this group of patients separately due to the possibility of inadequate power. The majority also have hypertension and not the rest of the comorbidities (64.2%). The Venn diagram can also be used to explain the concept of mutually exclusive events and probability of events. It can be extended to more than three categories, but the interpretation can be unwieldy.

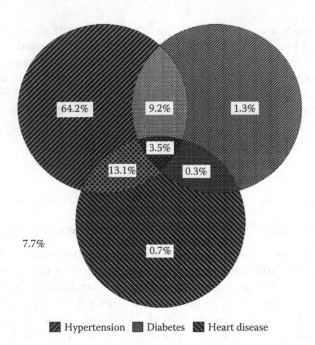

FIGURE 6.2
Venn diagram for patients with hypertension, diabetes, and heart disease.

6.4.2 Using the Gaussian Curve

The Gaussian curve (or inverted bell-shaped curve) is a useful figure to describe the normal distribution, which is a foundational assumption (i.e. normality in distribution of the variable) that needs to be met for many parametric tests, such as the independent Student's *t*-test or the linear regression model. Using a picture of the Gaussian curve, it may be easier to describe concepts such as mean and skewness in distribution of a continuous variable as well as the direction of skewness and the appropriateness of the use of the mean *vis-à-vis* the median in a skewed distribution. For instance, a different right-skewed distribution can be used to demonstrate how an outlier patient staying much longer than the rest of the patients has made an impact on the shape of the distribution. In addition, concepts in statistical process control such as the control limits based on the standard deviations, where for instance two standard deviations will contain approximately 95% of the observations (Figure 6.3) can be elaborated upon. For the clinician who is more mathematically inclined, the standard normal distribution can be used to demonstrate how probabilities are calculated, or even how variables measured on different scales can be standardised, combined and compared. Two normal distributions can be overlaid to demonstrate how hypothesis testing is carried out and the corresponding *p*-values calculated.

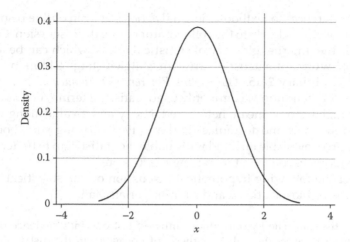

FIGURE 6.3
The standard normal distribution. *Note*: Vertical lines represent two standard deviations.

6.4.3 Simplifying Language

Wherever possible, simple language should be used to communicate the statistical analysis plan with clinicians. Statistical jargon and technical terms should be avoided wherever possible. For instance, instead of H_o and H_a which are typically used in the statistical literature, words such as the null hypothesis and the alternate hypothesis may be used, respectively. When examining the assumption for homogeneity of variance among groups in a linear regression model, it may be a good idea to use the phrases such as 'equality of variance' or 'homogeneity of variance' instead of the statistically heavily loaded word 'homoscedasticity!' Examples can also be used to describe statistical concepts, and these may not necessarily have to be about the medical field. For instance, in describing hypothesis testing, one may define the null hypothesis as 'Gender is not associated with height' or even use more catchy ones such as 'Students' school grades are not associated with their socio-economic status'. Some terms such as independent and dependent variables are often used interspersed with covariates and outcomes in the literature, but it would be better to standardise and use the same terms regularly in discussions with collaborators to avoid confusion.

One should also check with the collaborator regarding the person's background knowledge on basic statistical concepts such as hypothesis testing, p-values, confidence intervals and distribution theory, before making any reference to these terms during the conversation. In the event that the collaborator is not familiar with such terms, some reasonable amount of time needs to be spent in order to explain them clearly. For instance, it may not be appropriate for the statistician to quote the adjusted R-squared statistic from the linear regression model if the collaborator does not know what it means and signifies in the context of model selection. If necessary, the statistician can make references to textbooks, journal articles or even online resources to explain these terms clearly to the collaborator before the discussion. One such resource is the 'Internet glossary of statistical terms' which can be accessed here: http://www.animatedsoftware.com/statglos/statglos.htm#index. Date accessed: 2 February 2015. The *Collins Reference Dictionary of Statistics* also provides a compendium of commonly used statistical terms such as bimodal distribution, harmonic mean, heteroscedasticity and z-value along with the associated symbols and formulae (Porkess, 1988). Having such books and other resources available on hand will enable the statistician to better explain statistical terms to collaborators during discussions.

Consider the following hypothetical discussion on the statistical analysis plan between a biostatistician and a junior geriatrician:

> Biostatistician: The Kolomogonov-Smirnoff test provides evidence that we need to reject the null hypothesis of normality in the distribution of the 6 minute walk test (6MWT). Therefore we need to use a non-parametric test, specifically the Wilcoxon rank-sum test to test whether the distribution of 6MWT differs significantly among male and female stroke patients undergoing rehabilitation.

Although it may be appropriate to write the name of the normality test in the grant proposal or publication, it is not essential to describe this to the clinician during the discussions as he/she may not be familiar with them. Also, the word 'non-parametric' may not be easily understood by many clinicians. A simpler alternative description would be the following:

> Our main outcome measure is 6MWT, and we would like to compare this between males and females in the study. Since 6MWT does not appear to be normally distributed, we propose to use the alternative to the independent t-test, namely the Wilcoxon rank-sum test to calculate the p-value.

Here we assume that the clinician understands the basic concepts such as normality and *p*-values; otherwise, these need to be explained beforehand.

6.4.4 Employing Drawings

Drawings can be used to exemplify statistical concepts in an effective manner. For example, in a post-linear regression modelling exercise, rather than trying to describe to the collaborator the concept of testing for the assumption of 'homogeneity of variance of residuals', one may wish to draw a figure as shown in Figure 6.4. Counter-examples can easily be drawn to highlight possible violations to this assumption. Similarly, the concept of testing for interaction and demonstrating a difference in slopes in a linear regression model can be better demonstrated with the aid of a drawing. As we can see from Figure 6.5, a simple drawing can be used to highlight a difference in

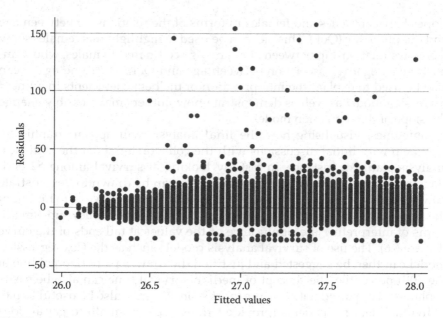

FIGURE 6.4
Explaining the homogeneity of variance of residuals assumption.

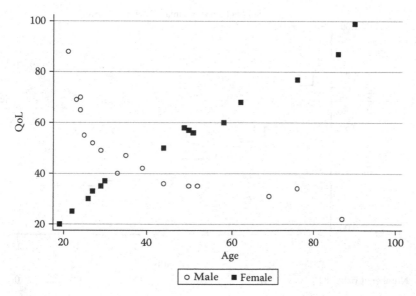

FIGURE 6.5
Explaining the interacting effect of age and gender on QoL.

slopes between males and females in terms of the relationship between age and quality of life (QoL). This plot can be used to highlight that females show a positive relationship between QoL and age compared to males, who demonstrate a negative association between age and QoL. The same figure can also be used to explain the interpretation of the beta coefficients in a linear regression model as well as demonstrate how outliers may possibly change the slope of the regression model.

Sometimes, visualising how the final analysis will appear graphically may help with initial discussions with the collaborator about the statistical analysis plan. For instance, in a study looking at survival among 82 end-stage renal disease patients, a survival curve can be drawn to demonstrate how censoring occurs when there is a step-down in the curve or to highlight the importance of including numbers at risk across the follow-up time in terms of interpreting the significance of the values at tail ends of the curve (Figure 6.6). The use of survival analysis models such as the Cox regression model can then be suggested and justified. By drawing a horizontal line at the 50th percentile, the concept of median survival time can also be easily explained. For paired data (e.g. pre/post scores), it may also be useful to plot individual data for patients (provided the sample is small) to get an idea of the change in scores then to purely look at the average value across all patients, as the latter can sometimes be misleading (Figure 6.7). Such plots can also be used to explain to collaborators the necessity for repeated measures analysis or random effects modelling.

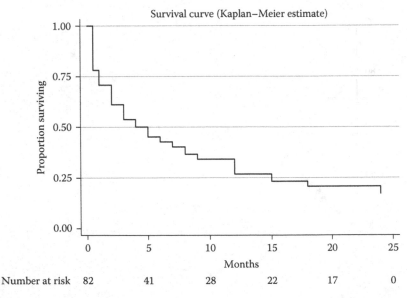

FIGURE 6.6
Explaining a survival curve, censoring and numbers at risk.

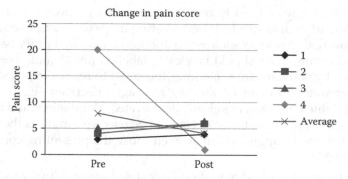

FIGURE 6.7
Plotting pre/post values for paired observations.

6.5 Effective Presentation Skills

For most part of his/her work, the statistician is often seen in front of the computer, burrowing away on an analysis of a dataset or writing up the statistical result for a manuscript in fervour. Interaction is usually with collaborators, discussing about collaborative projects or with fellow statisticians conversing about new statistical methodologies or challenges in data analysis. Occasionally, he/she may get an opportunity to present research in front of an audience. If the research is his/her own (e.g. a methodological paper or research project where he/she is the principal investigator, or work leading towards a masters or PhD thesis), the work may then be presented at a statistical or related conference or an oral viva defence in front of an exam panel. In addition, the person may be asked to present statistical methodology work to an interest group or a journal club. Some statisticians also may get invited to speak at seminars or courses. It is important then for the person to be familiar with some of the principles of good presentation and also learn PowerPoint tips that can make their presentation more effective and efficient. This section will cover some of the tips and tricks in PowerPoint that is customised for a statistical presentation as well as useful guidelines for the actual presentation and some points to note when creating a poster for presentation.

6.5.1 Tips and Tricks within Microsoft PowerPoint for a Statistical Presentation

1. As a rule of thumb, there should be one slide for every minute of presentation. Each slide should have 8–10 lines, clearly written words with at least a font size of 24, which should be seen from the back of the room. In particular, if equations are inserted in the text, ensure that they are of compatible font and size.

2. Graphics may be used in presentations as they convey much more information than words. When creating maps or charts, one should be aware that colours such as red and green usually signify bad and good events. Axes should be clearly labelled (the default axes provided by software are sometimes too small to read). Graphs should be annotated to convey the key message effectively (e.g. outliers highlighted or subgroups clearly demarcated). Animations and videos can captivate the audience but should be used sparingly as they tend to slow the computer and may even cause the presenting computer to crash!

3. For tables, the title should be clearly stated along with footnotes to identify the statistical tests used to calculate p-values. Numbers and p-values should be presented to two significant digits. Large numbers should be rounded (e.g. 7,000,000 displayed as 7 million). Tables should not be cluttered or busy, and ideally have a maximum of 20 rows and 4 columns of data displayed. For trend data, a graph is preferable to a table. Sometimes, a graph can be supplemented with a table to maximise information displayed (e.g. showing deaths and numbers at risk below a survival table).

4. One should not cram too many tables or figures in the presentation. Only the key results relating to the hypothesis being tested should be highlighted and discussed. In addition, reading every single result from a table or figure is not a good idea as this will bore the audience.

5. Slide animation may be used, but excessive variation in the type of animation and timing of slides should be minimised as they can be distracting. Slide numbers may be useful for both the presenter and the audience to gauge the number of slides remaining in the presentation.

6. Additional slides that are not be material to the presentation, but which may help to address any potential questions (e.g. simulation studies, literature review), can be appended at the end of the presentation or even hidden. These can then be refreshed whenever relevant questions are raised during the presentation.

7. Depending on the type of audience and nature of presentation, statistical formulae and equations may be presented. All symbols and statistical notions displayed need to be defined and adequate time should be allocated to discuss them during the presentation. Special care is needed when presenting any subscripts, superscripts and italics in the equations.

8. Always check the presentation for spelling and grammar mistakes or the presentation may appear sloppy or hastily put together. This is particularly so when most spell-check programmes may not recognise statistical phrases and terms.

6.5.2 Actual Presentation

1. Always make sure that you refer to your statistical symbols correctly. Mistakenly calling sigma (σ) for delta (δ) may just confuse your audience. Similarly, refer to the correct subscripts in the equations as these can sometimes be difficult to read.

2. Ensure that your slide contents are matched to the profile of the audience. If the majority of the audience are clinicians, there is no point in showing slides that derive the likelihood function and the process of maximising the likelihood. Similarly, if the crowd is predominantly statisticians, more details on model building, simulation analysis and computer codes may be appreciated.

3. Hand gestures can be used to emphasise key statistical concepts (e.g. draw a normal distribution or explain median by ordering data from the smallest to the largest value from left to right).

4. Avoid reading from the slides word for word. Similarly, do not read out every single result from the tables. Summarise key points or make some notes about the key message you wish to convey. Also ensure you do not rush through the slides and have enough time to describe them adequately.

5. Engage your audience by encouraging them to ask questions or even asking them questions. Try to weave in a story or relate an anecdote during your presentation. Give examples when describing statistical concepts (e.g. confounding or interaction). It also helps to get feedback from your audience after the presentation. This will help improve subsequent presentations.

6. Monitor the reaction of your audience and adjust your presentation as you go along. This is particularly when you are presenting 'theoretical statistical' stuff. For instance, if the audience displays a puzzled look on their face, read through the slide carefully as there could be a mistake in it. If the audience looks bored, you may wish to move on to other slides or keep the presentation brief.

7. Always remember to bring a backup copy of your presentation slides on a thumb drive, and/or in your e-mail, that can be retrieved in case the presentation file (and the backup file!) is lost.

8. Practice, practice, practice. Always remember to rehearse, as practice will only make your presentation get better. If possible, practice in front of your statistical colleagues and obtain their feedback.

6.5.3 Preparing a Poster

The checklist in Section 6.5.4 was adapted from a list in an article by J.E. Miller, (2007). Where possible, statistical examples have been provided for each point.

6.5.4 Checklist for Preparing and Presenting an Effective Research Poster

6.5.4.1 Content

- Design poster to focus on two or three key points. The audience's attention is lost if there are too many aims and hypotheses. Also, the entire poster will be very long, especially with methods and results to describe for the many aims. Usually one main aim along with perhaps another two secondary aims should be sufficient. At present, the most important points in the top two-thirds of the poster as readers are less likely to look at the bottom portion of the poster.

- Present materials that suit the background of the audience. For example, provide equations and their explanations for a statistical audience, and text descriptions of models for a clinical audience. Clinicians may also be more interested to read about clinical applications rather than simulation studies. When presenting equations, try not to copy and paste equations from other sources, but rather try and create them with the in-built equation editor.

- Paraphrase descriptions of complex statistical methods. Avoid excessive technical jargon and display of derivation of techniques unless that is the main focus of the research. For example, it may not be necessary to show how the likelihood function of the spatio-temporal random effects Bayesian model was derived if the focus of the poster is to highlight the application of the model to study out-of-hospital cardiac arrests.

- Spell out acronyms (e.g. autoregressive integrated moving average for ARIMA models or analysis of covariance for ANCOVA). Never assume that the audience is familiar with them even if they are statisticians.

- Replace large detailed tables with charts or small, simplified tables. For instance, if there are a large number of clinical, demographic and hospital institution-level factors associated with a health outcome, it may be possible to separately present them into three separate tables.

- If possible, accompany tables or charts with bulleted annotations of major findings. For example, in a study presenting three different diagnostic biomarker tools to predict prostate cancer, three different receiver operating characteristic curves may be presented. It would be meaningful then to annotate the individual area under the curves (AUCs) and perhaps even highlight the corresponding p-values for comparing the AUC among the groups.

- Describe the direction and magnitude of associations in the results. Consider presenting only variables of interest in the multi-variate table and list the other unimportant ones in the footnote. Instead of reporting that 'Diabetes was associated with poorer quality of life', it is better to report that 'Diabetes was *positively* associated with poorer quality of life *with an odds ratio of 2.5'*.

- Use confidence intervals and *p*-values to denote statistical significance. There is no need to present test statistics or standard errors as these take up space and relatively not as important. Sometimes, when there is a space constraint, one may wish to use symbols to denote statistical significance (e.g. * or δ).

6.5.4.2 Layout and Format

- Organise the poster into the background, data and methods, results and study implications (or conclusions). Some conferences may provide specific templates/instructions to follow.
- Divide the material into vertical sections on the poster. This makes it easier for the reader to follow the text and figures.
- Use at least 14-point type in the body of your poster and at least 40-point type for the title.

6.5.4.3 Narrative Description

- Have a couple of sentences overview of your research objectives and main findings. This will be useful when someone stops by your poster.
- Write short modular descriptions of specific elements of the poster to choose among the following in response to viewers' questions:
  - Background
  - Summary of key studies and gaps in the existing literature
  - Data and methods
  - Each table, chart or set of bulleted results

Sometimes, posters are displayed in electronic forms. For instance, the American Statistical Association on Statistical Practice 2014 conference (http://www.amstat.org/meetings/csp/2014/postertips.cfm. Date accessed: 23 April 2015) allowed for electronic poster (e-poster) presentations, which are similar to traditional poster presentations. They are presented on a large 42" computer screen, which is provided along with a laptop for each presenter. Clear instructions on how to format the poster, including a standard PowerPoint presentation template, is also provided in the aforementioned website.

6.5.5 Some Tips on Creating a Poster in Microsoft PowerPoint

- Click on Design→Page Setup, and then click on 'Slides sized for' and choose Custom. Then select the appropriate Width and Height. An A0 size poster would measure 84.1 × 118.4 cm, whereas an A1 size is 59.4 × 84.1 cm.

- Consider having text in two to three columns and read from left to right, top to bottom. Under Home→Paragraph, click on 'Columns' and select the required number of columns.
- Avoid excessive use of colour contrast as what you print may be different from what you see on the screen.
- Ensure that when you resize images, charts or tables, they are kept in proportion. In addition, consider having margins of at least 1 inch for your poster, to avoid text/images being truncated.
- Ensure only good quality images are used in your presentation (at least 100 dpi).

More useful suggestions can be found in 'Creating Effective Poster Presentations' (http://www.ncsu.edu/project/posters/index.html. Date accessed: 6 February 2015). Posters also come in different sizes and formats. A list of ready-made poster templates can be accessed here: http://www.postersession. com/poster-templates.php. Date accessed: 23 April 2015.

6.6 Tools and Resources

1. A useful document that details how best to create tables and charts such as bar charts and population pyramids and when not to use charts and mapping tools. The United Nations Economic Commission for Europe. Making Data Meaningful Part 2: A guide to presenting statistics. http://www.unece.org/stats/publications/ publ_topic/dissemination_2.html. Date accessed: 3 February 2015.
2. A compendium of statistical resources, training courses and videos on common statistical terminologies, which aims to support statistical literacy development. Understanding Statistics. Australian Bureau of Statistics. http://www.abs.gov.au/websitedbs/a3121120. nsf/home/statistical+language. Date accessed: 3 February 2015.
3. An excellent resource that provides definitions and explanations of elementary statistical terms and tests, with diagrams to aid explanations where possible is David Nelson's. *The Penguin Dictionary of Statistics* (2004).
4. YouTube video on communicating results to a non-biostatistician. Biostat Tutorial: Communicating Results. https://www.youtube. com/watch?v=s5tV727P0Gc. Date accessed: 3 February 2015.
5. Excellent tips on creating PowerPoint slides for research. The Cambridge Centre for Health Services Research. The art of making PowerPoint look good. http://www.cchsr.iph.cam.ac.uk/1110. Date accessed: 3 February 2015.

6. Oral presentation advice for researchers which includes a conference talk outline as well as an interesting article on 'How to give a bad talk'. Dr Mark Hill. University of Wisconsin–Madison, Madison, Wisconsin. http://pages.cs.wisc.edu/~markhill/conference-talk.html. Date accessed: 3 February 2015.

7. Webinars (online seminars) on oral presentations, designing PowerPoint slides for scientists provided by the Principal Investigators Association™. https://principalinvestigators.org/product/how-to-prepare-an-award-winning-oral-presentation/?utm_source=dms&utm_medium=email&utm_campaign=P150629. Date accessed: 1 July 2015.

8. *Information Visualisation.* A useful book that describes a number of advanced graphics including presentation of trivariate data, mosaic plots and map visualisation tools (Spence, 2001).

9. *Graphical Exploratory Data Analysis.* This book contains 180 graphical representations, including an overview of the most well-known and widely used methods of analysing and portraying data graphically. The emphasis is on exploratory techniques with real-life data used. Examples include multi-dimensional scatter plots, graphs used in cluster analysis and multi-dimensional scaling, regression model and control charts (du Toit et al., 1986).

6.7 A Collaborative Case Study

In this section, an interview with an experienced environmental epidemiologist is presented, in order to better understand some of the positive aspects of communicating with a biostatistician as well as identify areas that can be improved.

Collaborator:

Associate Professor Geoff Morgan

School of Public Health, Sydney University

Geoff Morgan has more than 15 years of experience in environmental epidemiological research and environmental health policy. Since being awarded his PhD in 2002, he has built a strong record of research in many areas of environmental health through active involvement of numerous trans-disciplinary collaborative teams funded by nationally competitive grants. He has applied the state-of-the-art exposure assessment and statistical approaches for epidemiological studies (such as time series and case crossover analyses, geospatial techniques) to investigate environmental risk factors for health.

Collaborative projects with authors:

1. Earnest A, Beard JR, Morgan G, Lincoln D, Summerhayes R, Donoghue D, Dunn T, Muscatello D, Mengersen K. Small area estimation of sparse disease counts using shared component models – Application to birth defect registry data in New South Wales, Australia. *Health Place*. 2010 Jul;16(4):684–93. Epub 25 Feb 2010.

2. Beard JR, Lincoln D, Donoghue D, Taylor D, Summerhayes R, Dunn TM, Earnest A, Morgan G. Socioeconomic and maternal determinants of small for gestational age births: Patterns of increasing disparity. *Acta Obstetricia et Gynecologica Scandinavica* 2009 Mar 27:1–9.

3. Beard JR, Tomaska N, Earnest A, Summerhayes R, Morgan G. Influence of socioeconomic and cultural factors on rural health. *Aust J Rural Health*. 2009 Feb;17(1):10–5.

4. Earnest A, Morgan G, Mengersen K, Ryan L, Summerhayes R, Beard J. Evaluating the effect of neighbourhood weight matrices on smoothing properties of Conditional Autoregressive (CAR) models. *Int. Journal of Health Geography* 2007 Nov 29;6(1):54–57.

5. Beard JR, Earnest A, Morgan G, Chan H, Summerhayes R, Dunn TM, Tomaska N, Ryan L. Spatio-temporal analysis of the relationship between socioeconomic disadvantage and admissions, associated procedures and mortality from acute coronary events in New South Wales, Australia. *Epidemiology*. 2008 May;19(3):485–92.

Date of interview: 1 August 2015

Q1. In the past 10 years, how many biostatisticians have you worked with?

Being an epidemiologist, I always try to work closely with at least one biostatistician on a regular basis. I am involved in various collaborative projects with different research groups and 'always' recommend that each project has an identified biostatistician. I have worked with at least six biostatisticians on various projects in the past 10 years.

Q2. For each project, how often do you meet up and discuss with the statistician?

It varies depending on the scope of the project and my role in the project, but we always have at least one longer meeting at the start of the project to scope out the study design/analysis/methods issues and the roles/responsibilities and timelines of all those involved in the project/analysis.

Q3. Are there any challenges with discussing statistical or methodological issues with the statistician via Skype or other electronic media *vis-à-vis* face-to-face meetings?

It continues to surprise/disappoint me how unreliable/clunky remote communication is in our Internet-connected world. Meetings with one other person via phone/Skype and sharing computer screens or e-mailed materials are generally fine, but Internet communications with more than one person simultaneously can still be unreliable. In this situation, I always try to have a backup phone conference number that can be called by the individual/group if the Internet-based communications fail, and try to e-mail copies of important materials before the meeting rather than rely on the online technology.

Q4. What has been the most difficult to understand the statistical phrase or concept communicated to you?

The association is not significant.

Q5. What are some of the challenges in communicating with biostatisticians?

Tailoring the detail of the analysis to fit the study design/data quality/target publication issues. Any specific statistical analysis can be conducted using a number of different approaches, and the art is to identify the approach that provides a ROBUST answer that the collaborators are confident in without unnecessary/redundant analysis or applying technical/methodological analyses that provide meaningfully equivalent results to more simple and 'understandable' analyses.

Q6. Can you suggest some ideas that can enhance or improve communication between statisticians and their collaborators?

   i. Use lay terms with non-statistical collaborators whenever possible. Do not underestimate the value/input of colleagues just because they are not able to communicate in statistical terminology/concepts.

  ii. Clearly define the role of each within the context of the specific collaboration.

 iii. Clearly identify the resources available for the project, including the analysis and write-up. This can help match the scope of the analysis to the available resources for the project and match the level of statistical detail required for the analysis and write-up.

 iv. Clearly state the main hypothesis for the study.

  v. For regression-type analyses, write down the conceptual model for the analysis. This can be a simple regression equation that defines the health outcome, covariates and study factor (e.g. exposure) of primary interest. This is essentially a summary of the study in a symbolic form that can help focus on the statistical model that will need to be implemented.

vi. If a project is worth doing, it is worth documenting so that it can be redone by someone else using the same data at some future time. This includes defining study design, data, analyses and associated information, including appropriate meta data. Everything should be done with the thought that it may need to be redone at some point (e.g. in 6 months/1 year time due to reviewer comments). Data/file management and documentation is an essential part of any collaborative project – not a *post hoc* afterthought.

Key Learning Points

1. Ensure that you are well prepared for the initial meeting with the collaborator.
2. Acquire skills in deciphering medical terminology.
3. Communicate statistical phrases more effectively to your collaborators.
4. Obtain useful tips on creating PowerPoint slides to effectively communicate your research.
5. Learn beneficial presentation skills for presenting statistical research.
6. Find out how to prepare a statistical poster for presentation.

References

du Toit, S. H. C., Steyn, A. G. W., & Stempf, R. H. (1986). *Graphical Exploratory Data Analysis*. New York: Springer-Verlag.
Miller, J.E., 2007, Preparing and presenting effective research posters, *Health Serv Res Jou*, 42(1), 311–328.
Porkess, R. (1988). *Dictionary of Statistics* (1st ed.). Glasgow: Collins.
Spence, R. (2001). *Information Visualization*. Essex: ACM Press.

7

Effective Writing Skills

7.1 Introduction

Although statisticians are undoubtedly good with numbers, some may find it challenging when it comes to contributing towards the writing up of the results for the manuscript or report. The problem may be more severe for those for whom English is not the main language of education. In addition, the specific type of writing contribution differs according to the nature of the report (i.e. whether it is a grant application, a publication or even a report for a hospital/health care institution), and the strategy for tackling the different elements within the document may not be obvious. Although resources on improving general writing skills are available, none are targeted for a biostatistical collaboration. In this chapter, we provide suggestions on how the writing process can be made easier for the statistician by providing a structure and guideline on how to write the various sections of the grant proposal, publication or report. A distinction is also made between writing for a first author versus a collaborative publication. Established international guidelines for the various epidemiological study designs will be used to demonstrate how sections within a manuscript or grant proposal can be written. Topics such as selecting the appropriate journal and following the instructions for authors in the journals will be discussed with relevant clinical examples.

7.2 Writing a Statistical Analysis Plan for a Grant Application

Study design and statistical analysis plans are important considerations in a grant proposal. Most grant funding agencies explicitly state detailed requirements for these in their application forms. In addition, some like the National Medical Research Council (NMRC) in Singapore provide checklists in its application guide for researchers to follow when they apply for a grant (National Medical Research Council, 2015), such as the following:

1. Is the main objective exploratory (for which a formal sample size justification is not relevant) or are you testing a quantitative hypothesis?
2. If the former, what are the population parameter(s) you are trying to estimate? (e.g. annual incidence of acquired immunodeficiency syndrome, prevalence of teenage smokers, relative risk of a relapse, etc.)
3. If the latter, use the checklist below which could assist you to describe your study design and methods more clearly. Note that the checklist provided by the NMRC is adapted from the Consolidated Standards of Reporting Trials (CONSORT) statement, and this will be discussed in more details later in this chapter.
4. Some sample questions related to the checklist below: (a) State the primary quantitative hypothesis (e.g. the hazard ratio is 0.5) and the required precision for the estimate in the form of a confidence interval (CI), for example, 95% CI of the hazard ratio (HR; 0.3, 0.7)? (b) State the sample size required to achieve that precision. Be sure to specify the type 1 error, any other required assumptions needed as well as the sample size software/formula used in the calculation.

Usually, grant authorities specify a page limit for the research proposal, and also provide structured subsections for investigators to fill up. This not only helps investigators to limit the writing into a succinct document but also aids with the peer-review process. The NMRC allows a maximum of 12 pages (excluding references) for the research project and has the following subsections shown below (National Medical Research Council, 2015). This may differ slightly across various grant awarding authorities (e.g. the National Health and Medical Research Council has a nine-page limit for research proposals), but the basic information required remains pretty much the same:

- Specific aims and hypothesis
- Background and clinical significance
- Preliminary studies/progress report
- Methods/approach
- How the research furthers the vision/mission of NMRC
- Roles of team members
- References

The statistician's input is traditionally sought for the "Methods/Approach" section, but it is important that he/she provides critical review for the other relevant sections as well.

7.2.1 Specific Aims and Hypotheses

Although it is the principal investigator (PI)'s responsibility to write out the study aims, the statistician should be closely involved to ensure that the aims can be translated into hypotheses that are readily testable and that the results are achievable in terms of a realistic sample size. One of the main difficulties in formulating the research aim relates to clarity, specifically in terms of trying to summarise a long statement or even paragraph into a brief sentence. In the proposal, it is a good idea to limit the main objective to one, secondary objectives to three and exploratory aims to two. These aims should obviously be listed in decreasing order of importance. The primary objective is important as it will form the main justification for the study, and sample size/power calculations will also be based on this. Reviewers also tend to pay particular attention to this section.

Each aim in the study should be linked to a specific hypothesis. The aim basically states the question to be answered, whereas the hypothesis specifies the operational elements necessary and essential in the particular project, which will be used to answer the question. For example, an aim would be to study whether a multidisciplinary team case management approach (MTCM) will help reduce the length of stay (LOS) in elderly lung cancer patients compared to usual treatment, as described in the example below. Sometimes, clinicians have problems at this stage and are unable to write down a hypothesis that is specific enough. The statistician should work closely with the clinician to ensure that a workable hypothesis is derived with the WHO (subjects, raters or treaters), HOW (by what means or with what measurements will be taken), WHAT (the 'thing' being studied), WHERE (site/setting), WHEN (in the course of treatment, how long), and BY WHAT METRIC (what is the measure, effect size) included in the hypothesis described previously in Chapter 1 (see Section 1.4).

Let us look at a few illustrations.

Example 1

> *Aim.* To study the effectiveness of a MTCM approach to reduce LOS among elderly lung cancer patients in the hospital.
> *Initial hypothesis.* Elderly patients with lung cancer who are assigned to MTCM would stay for a lesser period in hospital compared to those who were not. This hypothesis is not specific enough and does not state the effect size (i.e. what is the minimum clinically important difference in LOS that the investigators wish to detect between the two groups?). This would in turn help with subsequent sample size calculations. The study period should also be stated to give an idea of timeline and feasibility of sample recruitment, as well as some indication of how the data were collected (measurement) to assess reproducibility.
> *Improved hypothesis.* Elderly patients with lung cancer who are admitted from 2014 to 2015 to hospital X and assigned to MTCM

on admission would stay on average 2 days lesser compared to those who are not, as recorded from the hospital inpatient admission system.

Example 2

Aim. To investigate whether patients admitted with human immunodeficiency virus (HIV) suffer greater impairment in quality of life (QoL) then patients admitted to the hospital with dengue infection in Melbourne.

Initial hypothesis. Patients admitted for HIV will have a lower QoL score compared to those with dengue infections. This hypothesis is lacking in a number of details. We do not know the instrument that is used to measure the QoL, the person(s) administering the tool, at which stage of the admission period the tool will be administered, along with the effect size of interest.

Improved hypothesis. Patients admitted for HIV in the hospital from January to December 2010 will have a lower QoL score (i.e. 5 units lesser) measured on the day of admission by a trained nurse using the SF-36 instrument, compared to those with dengue infections.

The way the hypothesis is written will have a profound effect on the specification of the statistical test to be performed in the statistical analysis plan and the way the statistical approach is written and described. This is exemplified with an example on high cholesterol levels and diagnosis of stroke, as highlighted in Table 7.1. If the hypothesis is written as 'cholesterol levels (outcome) are associated with diagnosis of stroke', then an independent Student's *t*-test may be required, whereas if the hypothesis is written as 'high cholesterol levels (i.e. >6.5 mmol/L) is associated with stroke, regardless of age', then the logistic regression model would

TABLE 7.1

Relationship between Hypothesis Statement and Statistical Test

Hypothesis	Statistical Test/Analysis Plan
Cholesterol levels are associated with diagnosis of stroke.	Independent Student's *t*-test/linear regression model
Cholesterol levels are associated with diagnosis of stroke, independent of age.	Linear regression model, with adjustment for age
The effect of cholesterol on diagnosis of stroke is higher for men compared with women.	Binary logistic regression model, with interaction between gender and cholesterol
High cholesterol level (i.e. >6.5 mmol/L) is associated with stroke.	Chi-squared test/binary logistic regression
High cholesterol (>6.5 mmol/L) is associated with stroke, regardless of age.	Binary logistic regression, with adjustment for age

be applicable. In particular, the nature as well as the number of variables in the hypothesis would dictate the type of statistical test required.

7.2.2 Background and Clinical Significance

This section is mainly for the PI to fill up. If there are numbers being presented (e.g. burden of disease, hospitalisation costs, etc.), the statistician can help to ensure that the appropriate statistics are presented (e.g. median if data are skewed/expected to be skewed). The statistician should also help ensure that the gap identified in literature is linked to the subsequent aims and hypotheses listed in the project. Sometimes, the background is written initially and then the aims/hypotheses are significantly revised following subsequent discussions with biostatisticians. In that case, it is important to ensure that the background of the study is amended accordingly.

7.2.3 Preliminary Studies/Progress Report

This section is important as it allows reviewers to assess whether the project is feasible and whether the PI and the rest of the study team have demonstrated the ability to work together and see the project through. Results from any previous/pilot study should be presented. Important statistics such as dropout rates, estimated effect size, prevalence of exposure and outcome as well as means and their estimates of variability in continuous outcome measures should be reported. This information can subsequently be used in the planning of the study design and sample size considerations for the current proposal. Potential problems identified in the pilot study (e.g. high dropout rates and low prevalence of outcomes), as well as how the current study aims to address those limitations, should be highlighted. For instance, if the pilot study indicated a slow recruitment rate, amendments to the current proposal to address the limitations should be highlighted (e.g. lengthening the study period or conducting multi-centre trials). Data may be summarised in tables and figures (preferably kept to a maximum of 1), and it is only necessary to present the main results in relation to the current grant application. If the pilot study is similar to the main proposal, it must be made clear that the main study will not include data from the pilot in the analysis.

7.2.4 Methods/Approach

This is one of the most important sections of the grant, where the statistician provides critical input. In general, the statistician is expected to undertake and report sample size calculations and write out the statistical analysis plan. However, depending on the complexity and nature of the study, various other inputs such as addressing measurement error or bias, treatment of missing data and randomisation codes and procedures may also be required. Usually, this section will have sub-headers, such as recruitment methods, interventions, outcomes, sample size and statistical analysis, and they may differ according to the study design. There are various checklists devised for different study designs such as randomised controlled trials (RCTs), cohort and case–control studies, and these will be discussed in Sections 7.2.4.1 through 7.2.4.4. Although these checklists were written primarily to help researchers report results from a study, certain parts can be used as templates for subsections in a grant proposal as will be demonstrated next.

7.2.4.1 CONSORT Statement

In an attempt to strengthen the reporting of results from RCTs, a group of trialists, journal editors and methodologists have since 1996 published a series of recommendations and checklists known as the CONSORT statement. There have been a number of revisions over the years, and the most recent one is known as the CONSORT 2010 checklist, comprising 25 items organised in sections, including introduction, methods, results and discussion (Moher et al., 2010). The CONSORT statement has been adopted by more than 400 journals from the International Committee of Medical Journal Editors (ICMJE) and has shown to improve the quality of RCTs reported (Moher et al., 2010).

If the grant proposal involves an RCT, then the CONSORT statement provides a template that is comprehensive and may significantly enhance the quality of information provided to the peer-review panel who will be evaluating the proposal. Depending on the nature of the trial, not all the 25 items on the checklist need to be discussed. For instance, if the trial is not blinded and there are no early stopping rules, then obviously these need not be discussed in the proposal. Registration, availability of protocol and funding may also be omitted from the proposal. It is not possible to discuss all 25 items of the CONSORT statement here, and readers can obtain the entire checklist and more information from the CONSORT website (http://www.consort-statement.org/. Date accessed: 11 February 2015). However, we summarise the important points on two of the main items in the checklist here (Moher et al., 2010):

Item 7 of checklist – sample size
- Identify the primary outcome on which calculation was based, along with all the quantities used in the calculation.
- Report the sample size per group and whether any allowance was made for attrition or non-compliance.
- Report whether there will be interim analyses performed, and if so, pre-specify any group sequential statistical methods that will be used to adjust for multiple analyses.

Item 12 of checklist – statistical methods
- Describe statistical methods in enough details to ensure reproducibility, intention to treat, missing data imputation and so on.

- Provide an estimate of effect size along with the CI.
- Provide actual p-values, rather than evaluating them against the standard threshold (i.e. results will be presented as $p < .005$ or $p < .001$).
- Be aware of the possibility of non-independence of observations in the data, and if necessary describe statistical methods to address the issue.

Example 3

Aim. To study the independent effects of a cognitive and exercise intervention in reducing the rate of neurodegeneration in pre-Huntington's disease patients.

Suggested writing

Sample size. The sample size was informed by a power analysis on the caudate (primary outcome measure). Based on a previous study, we estimated a difference of at least 0.012 (standard deviation [SD] = 0.02) in volume loss between the intervention and control groups. Using an alpha of 0.05, and with power of 0.80, $n = 45$ per group and assuming that 10% will drop out of the study (attrition rate estimated from the pilot study), we would then need 50 subjects per arm. We will then recruit a total of $n = 300$ close to onset (<10 years) pre-Huntington's disease participants (150 at each of two sites) who will be randomly assigned to one of three treatment arms: cognitive training (CT), exercise training (ET) or active control (AC). Combined across sites, there will be a total of $n = 100$ in each of the three groups ($n = 50$ per arm per site). Sample size was calculated using Power and Sample Size Program (V3.0).

Statistical analysis. To assess the efficacy of the cognitive and exercise interventions over the three time points, we will conduct linear mixed-model analysis. Separate models comparing CT and ET to the AC group will include the main effects for Group and Time and a Group x Time interaction term. Each outcome measure will be tested separately. Analyses will be intention to treat using multiple imputations for missing outcome data. Models will include age, sex, education, intelligence quotient (IQ), mental and physical activity levels and site as covariates. Outcome measures will comprise brain volume voxel-wise measures. Data analysis will be performed in Stata V13.0 (StataCorp, College Station, Texas) and the level of significance set at 5%.

7.2.4.2 STROBE Statement

The CONSORT statement was designed for RCTs. Much of biomedical research is observational in nature, and some of the most common types of study designs include cohort, case–control and cross-sectional studies.

The Strengthening the Reporting of Observational Studies in Epidemiology (STROBE) statement consists of a checklist with 22 items, including study design, setting, data sources, results and discussion (von Elm et al., 2008). It includes features that are unique to a particular study design (e.g. how was loss to follow-up addressed in a cohort study, how was matching performed in a matched case–control study, etc.). A full list of the checklist, including modifications for conference proceedings, guidelines for publishing in *PLoS Medicine* journals and extension of the checklist for reporting molecular epidemiology infectious disease studies can be accessed here: http://www. strobe-statement.org/index.php?id=strobe-home (Date accessed: 12 February 2015). We provide examples of how the checklist can be used to write the methods section of a grant that involves an observational study design.

Example 4

Aim. To identify specific demographic and clinical factors that are positively associated with mortality in patients with chronic obstructive pulmonary disease (COPD) using a cohort study design.

Sections 4, 5 and 6 in the STROBE statement relate to describing the study design, setting and participants, and state the following elements:

- Present the key elements of study design early in the paper.
- Describe the setting, locations and relevant dates, including periods of recruitment, exposure, follow-up and data collection.
- Give the eligibility criteria as well as the sources and methods of selection of participants. Describe methods of follow-up.
- Cohort study – For matched studies, give matching criteria and number of exposed and unexposed.

Suggested writing

We propose a cohort study design with the cohort defined as all patients diagnosed with COPD from 2014 to 2015 in hospital X, and followed up till 2020 or until mortality is observed, whichever occurs first. Diagnosis of COPD will be based on a spirometry test performed by a respiratory clinician. Exposure variables such as age at diagnosis, number of cigarettes in a typical week, lung function capacity and prevalence of comorbidities will be collected at the time of diagnosis. Patients will be followed up during their regular hospital appointments, and mortality will be recorded via the national birth and death registry. A dedicated trained nurse clinician will collect and update the data.

Example 5

Aim. To examine and quantify the association of *BRCA1* or *BRCA2* genetic mutation with breast cancer after adjusting for smoking using an age-matched case–control study. The secondary aim is to establish if there are any differences across ethnic groups.

Section 12 of the STROBE checklist relates to the statistical method and has the following suggestions:

- Describe all statistical methods, including those used to control for confounding.
- Describe any methods used to examine subgroups and interactions.
- Explain how missing data were addressed.
- Case–control study – If applicable, explain how matching of cases and controls was addressed.
- Describe any sensitivity analyses.

Suggested writing

In order to determine the association of *BRCA1* or *BRCA2* genetic mutation with breast cancer, we propose to undertake a multi-variate analysis with presence of breast cancer as the main outcome of interest. Prevalence of *BRCA1* or *BRCA2* will be the exposure variable, and current smoking status of the subject (coded yes/no) will be the confounding variable that will be adjusted for in the analysis. Smoking status variable will be forced in the multi-variate analysis and not subject to any form of stepwise model building. Because the design involves a 1:1 age-matched case–control study design, we propose to use the conditional binary logistic regression model in the analysis. In order to ascertain whether there are subgroup differences, we would test for an interaction between ethnicity and presence of *BRCA1/BRCA2* in the model. In the event that the interaction term is significant, we will present the results separately stratified by ethnicity. We do not expect substantial missing data (i.e. less than 5%), and thus propose to use all available data in the analysis. Missing data imputation techniques will not be undertaken. Data analysis will be performed in Stata V13.0 (StataCorp) and level of significance set at 0.05.

7.2.4.3 PRISMA Statement

The Preferred Reporting Items for Systematic Reviews and Meta-Analyses (PRISMA) statement started from an initial set of guidelines designed for meta-analysis, the Quality of Reporting of Meta-Analyses (QUOROM) statement, but has now expanded to systematic reviews. There is a 27-item checklist that aims to help authors improve their reporting of systematic reviews and meta-analyses (Moher et al., 2009). The entire checklist and the explanations can be obtained from the following website: http://www.prisma-statement.org/ (Date accessed: 12 February 2015). The following three items under the methods section of the checklist are discussed here:

Item 7 of checklist – information sources. Describe all information sources (e.g. databases with dates of coverage, contact with study authors to identify additional studies) in the search and date last searched. The search strategy should also include the review of the citations in the relevant studies or citation services such as the Science Citation Index as well as identifying the PIs who are leaders in the field, and thereafter searching their website for potential studies that can be included. Disease registries are also an invaluable source on research (either published or ongoing). For example, the Swedish twin registry currently has around 30 ongoing projects, covering a wide range of public health issues, including dementia, allergies, cancer and cardiovascular diseases (http://ki.se/en/research/about-us-the-swedish-twin-registry. Date accessed: 21 April 2015).

Item 8 of checklist – search. Present full electronic search strategy for at least one database, including any limits used, such that the search can be repeated. It is important that the systematic review is able to search and identify all the relevant studies on a particular topic, which can then be subsequently included in a meta-analysis. This is to ensure that the results are representative and also to minimise selection bias. The key strategy is to search for studies thoroughly and in a systematic manner. Index terms or phrases used to search online databases such as MEDLINE and EMBASE should be clearly described in the systematic review, along with the study dates keyed in and information provided on whether the people conducting the research have any prior experience or training in conducting systematic reviews.

Item 9 of checklist – study selection. State the process for selecting studies (i.e. conditions for screening, eligibility criteria, inclusion in systematic review and, if applicable, inclusion criteria for the subsequent meta-analysis). Key issues that should be discussed include whether unpublished and reports which are not peer-reviewed such as master's thesis and conference proceedings should be included in the review. Another important point to note is that many of the larger trials and cohort studies often result in multiple publications, and it is important to differentiate and include only one study in the review.

As meta-analyses are based on published papers, there is a concern that non-significant studies may not get published and hence not have an opportunity to be included in the meta-analysis study. This is known as publication bias, and funnel plots have traditionally been used to identify possible publication bias (Egger et al., 1997). It may not be advisable to combine results from unpublished studies with those that have been published as the former may not have gone through the same rigorous process of peer-review and appraisal. In the presence of publication bias, one may opt to do a stratified analysis among studies with large sample sizes with a view that these studies are unlikely to be affected by publication bias.

Example 6

Aim. To combine results from several studies to evaluate the diagnostic performance of a new biomarker in staging colorectal cancer patients.

Suggested writing

Statistical analysis. The accuracy of the new biomarker test in the staging of colorectal cancer patients diagnosed at the clinic will be determined by the pooled diagnostic odds ratio (DOR) and the summary receiver operating characteristic (SROC) curve. The DOR combines the effects of sensitivity and specificity into a single number, whereas the SROC provides an optimal threshold for the trade-off between sensitivity and specificity. Heterogeneity of the included studies will be explored from the forest plots, and more specifically established using the test for heterogeneity. In the presence of heterogeneity, the random effects method will be used to synthesise estimates of the effect size from individual studies. When significant heterogeneity is observed ($p < .05$), a random effects model will be applied. Subgroup analysis will be performed according to studies that are grouped into quartiles in terms of their study quality scores. Data analysis will be performed in Stata (StataCorp), and the level of significance set at 0.05.

7.2.4.4 STARD Statement

Similar to the checklists for the other study designs, the Standards for the Reporting of Diagnostic Accuracy Studies (STARD) was developed to improve the accuracy and completeness of reporting of studies of diagnostic accuracy, allowing readers to assess the potential for bias in the study (internal validity) and to evaluate its generalisability (external validity) (http://www.stard-statement.org/. Date accessed: 16 February 2015). The STARD checklist consists of 25 items, including describing the reference standard and its rationale and the technical specifications of material and methods. It also includes suggestions of providing information on how and when measurements were taken, citing references for index tests and reference standard, as well as describing definition of and rationale for the units, cut-offs and/or categories of the results of the index tests and the reference standard.

In particular, when reporting diagnostic studies, it is important to characterise the test results correctly. The usual statistics reported are sensitivity, specificity, positive and negative predictive values and positive and negative likelihood ratios. Sensitivity is the ability of the test to correctly identify subjects who have the disease or outcome. For example, when using a biomarker X as a screening tool to identify brain cancer verified via pathology samples, a resulting value of 90% sensitivity would mean that among all the samples identified by pathology results to be brain cancer, 90% of them would have a positive test for biomarker X. This is also known as true positives. Specificity is the ability of the test to correctly identify patients who do not have brain cancer. Eighty-five percent specificity would indicate that among all the non-cancer cases, 85% were correctly identified as not having the presence of bio-marker X. A positive predictive value of 75% means that of all the patients with a positive biomarker X reading, 75% of them will develop brain cancer. Similarly, a negative predictive value of 20% means that among those without biomarker X, 20% will not have brain cancer. The likelihood ratio combines both the sensitivity and specificity values, and a positive likelihood ratio of 2.5 indicates that a positive biomarker test result is 2.5 times more likely in a subject with brain cancer than in one who does not have brain cancer based on the pathology report.

Example 7

Aim. To determine whether inclusion of active asthma diagnosis improves diagnostic accuracy of severe COPD exacerbations (defined as more than three events in a year), compared to a model with just sex, age, smoking status and body mass index (BMI).

Suggested writing

Statistical analysis. The main outcome measure is severe COPD exacerbations (coded yes/no). Two separate logistic regression models will be run. The first model will include sex, age, smoking status and BMI as independent covariates. The second model will include the same variables as before, but this time with the addition of active asthma diagnosis. We will include all variables in the model, regardless of their statistical significance, as these were deemed clinically meaningful. Based on the final model coefficients, the predicted probability of outcome will be calculated for each patient under the two different models. ROC curves will used to calculate the area under the curve, which would then be compared across the two models, to measure their relative discriminatory properties. Based on the optimal cut-off points determined from the ROC curves, the associated sensitivities, specificities and positive and negative likelihood values will be calculated and used for clinical decision making. Data analysis will be performed in Stata V13.0 (StataCorp) and the level of significance set at 5%.

The Enhancing the Quality and Transparency of Health Research (EQUATOR) network also has a broad compilation of reporting guidelines for the main study types, including those we have already discussed as well as studies involving economic evaluations and animal pre-clinical studies (http://www.equator-network.org/toolkits/authors/. Date accessed: 6 April 2015).

7.3 Writing for a Publication – First-Author Publication

In Section 7.2, we discussed the writing of study design and statistical analysis plan in a grant proposal. Once the study has been completed and analysed, the statistician will usually be involved in writing up the methods, results and other sections of the manuscript. The level of input and commitment in the manuscript writing process would differ according to whether it is a first-author publication (usually statistical or methodological in nature) or a collaborative project, where the PI is usually someone else who will drive the entire writing process. In this section, we describe the writing process for a first-author publication, and in Section 7.4, writing a manuscript for a collaborative project will be discussed.

7.3.1 Selecting the Journal

Sometimes, a biostatistician, usually a senior faculty member, is in a position to write up a first-author publication. This may be a statistical methodology paper that is focussed on the development of a new model/methodology or an application of existing novel methods to clinical datasets or even comparison of different statistical models. Usually, such papers are published in methodology-focussed journals such as *Statistics in Medicine, Biostatistics, Statistical Methods in Medical Research* and *American Journal of Epidemiology*. The topics may be areas of specialisation for the statistician (e.g. work done during their PhD candidature or postdoc) or they can be motivated through the course of their work (i.e. developing novel methods of forecasting counts of diabetes at the geographical level when working with the local health department on geospatial mapping). The key element of such methodological research is that the statistician drives the research process and takes ownership of the entire project, from conceptualisation of the idea to the eventual submission of publication and follow-up revision/re-submission (if necessary). This is different from the input required in a collaborative project (discussed later). Some examples of both types of projects include:

1. Development of imputation methods for systematically missing predictor variables in individual participant meta-analysis (methodological)
2. Application of generalised estimating equations to study the impact of BMI on blood glucose levels (collaborative)

3. Comparison of the autoregressive integrated moving average model and the Knorr–Held two-component model to predict dengue fever notifications (methodological)

Before writing the article, it is a good idea to select the journal that one wishes to submit the article to in order to establish the writing style and structure that is required by the journal. Different journals have different scope and target audiences, and it is a good idea to visit their website and see if the research topic falls within the scope of the journal. For instance, the *Statistics in Medicine* website   (http://onlinelibrary.wiley.com/journal/10.1002/(ISSN)1097-0258/ homepage/ProductInformation.html. Date accessed: 19 February 2015) states that 'Papers will explain new methods and demonstrate their application, preferably through a substantive, real, motivating example or a comprehensive evaluation based on an illustrative example, and that papers with primarily mathematical content will be excluded'. Hence, it may not be a good idea to send a paper that is primarily theoretical or mathematical here. The *Annals of Statistics*, however, publishes statistical manuscripts that are often methodological (http://www.imstat.org/aos/policy1.html. Date accessed: 14 August 2015). Hence, studying the scope of a journal is a good idea, and it will help to better avoid unnecessary delays in the publication process arising from the manuscript not meeting the expected scope of topics within the journal and being rejected.

An equally important criterion in choosing the journal is the impact factor (IF). The general consensus is to submit the paper to a journal with a higher IF first, and then go down the list if the paper is rejected. For instance, *Statistics in Medicine* has listed its IF on its website as 2.04, whereas *Biostatistics* has a slightly higher IF of 2.24 (http://biostatistics.oxfordjournals.org/. Date accessed 19 February 2015). Another important consideration is whether there is an open access option for the author, where the paper is made freely available online immediately upon publication for a charge, although routinely the charges are waived or reduced for those submitting from developing countries. Sometimes, if the statistician has previously helped with the peer-review process for the journal, he/she may get a discounted price for publication, which may make the journal more appealing in terms of submission. Other important considerations include the target audience (i.e. statisticians vs. health policy makers) and even the statistician's familiarity or previous track record of publishing in a particular journal.

For the statistician who is publishing for the first time and needs suggestions on which journal to submit to, the Journal/Author Name Estimator (JANE) is a useful online tool (http://www.biosemantics.org/jane/. Date accessed: 19 February 2015). One needs to only key in the title or parts of the abstract in the search box, and the tool will provide suggestions on which journals to submit the paper to, as well as identify articles or authors with publications on the topic. For instance, when the phrase 'simulation model selection diagnostics cancer' was typed in JANE, the following journal names were suggested (Figure 7.1). The confidence bar on the left indicates the level of certainty of the match, and obviously the more phrases one uses, the better the match. Journals are also ranked by the 'article influence' index, which measures how often the articles in the journal are cited within the first 5 years of publication.

7.3.2 Instructions for Authors

Once the journal has been selected, the next thing to do is to read the 'instructions for authors' guideline of the journal. Most journals allow for an online submission, peer-review evaluation and eventual publication of an article. The instructions for authors vary across journals, and certainly the structure of the manuscript is different for a methodological paper, compared to medical journal articles (the latter will be discussed in Section 7.6). Therefore, it is important that the statistician has a look at the instructions before starting work on the manuscript. For instance, the *Statistics in Medicine* journal has author guidelines that cover manuscript style (including font size and type, inclusion of sponsors/grants, abstract), reference style, use of illustrations, supporting web materials and copyright and permissions (http://onlinelibrary.wiley.com/journal/10.1002/(ISSN)1097-0258/homepage/ForAuthors.html. Date accessed: 19 February 2015).

General suggestions across journals include using standard font types such as Times, Helvetica or Courier, using standard font sizes throughout the manuscript, avoiding unnecessary formatting such as bolding, italicing and underlining of texts, using graphics of good quality (at least 300 dots per inch [dpi]), avoiding colour figures and charts (some journals charge extra for the printing of colours), keeping figures and sometimes tables separate from the main text of the manuscript and so on.

Manuscripts are usually submitted via a LaTeX (a specialised manuscript formatting software) or a Microsoft Word document format. Mathematical equations can be easily included into the manuscript in Microsoft Word. For versions Word 2003 and earlier, click on 'Insert'→'Object' and then select Microsoft Equation 3. For Microsoft Word .docx documents, it is actually easier to insert equations. Just click on 'Insert'→'Equation', and one can then select from a range of built-in equations. The University of Waterloo, Waterloo, Ontario, Canada, has a useful website that provides handy tips on inserting equations, stacking and aligning equations and cross-referencing equations as well as useful instructional videos (https://uwaterloo.ca/information-systems-technology/services/scientific-computing-software-support/supported-software-scientific-computing/creating-numbering-and-cross-referencing-equations-microsoft. Date accessed: 5 March 2015). One should take note when aligning the equation to ensure that the size is similar to the rest of the text in the manuscript.

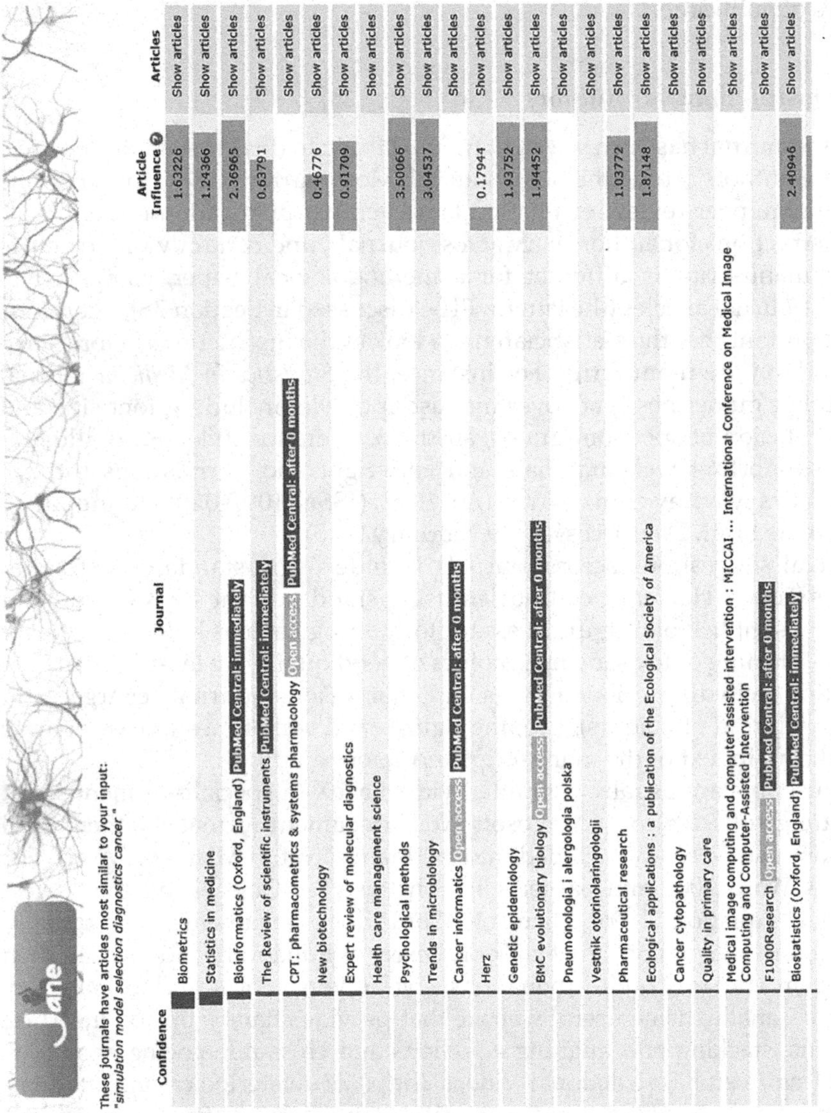

Jane

These journals have articles most similar to your input:
"*simulation model selection diagnostics cancer*"

Confidence	Journal	Article Influence ❓	Articles
	Biometrics	1.63226	Show articles
	Statistics in medicine	1.24366	Show articles
	Bioinformatics (Oxford, England) PubMed Central: immediately	2.36965	Show articles
	The Review of scientific instruments PubMed Central: immediately	0.63791	Show articles
	CPT: pharmacometrics & systems pharmacology Open access PubMed Central: after 0 months		Show articles
	New biotechnology	0.46776	Show articles
	Expert review of molecular diagnostics	0.91709	Show articles
	Health care management science		Show articles
	Psychological methods	3.50066	Show articles
	Trends in microbiology	3.04537	Show articles
	Cancer informatics Open access PubMed Central: after 0 months		Show articles
	Herz	0.17944	Show articles
	Genetic epidemiology	1.93752	Show articles
	BMC evolutionary biology Open access PubMed Central: after 0 months	1.94452	Show articles
	Pneumonologia i alergologia polska		Show articles
	Vestnik otorinolaringologii		Show articles
	Pharmaceutical research	1.03772	Show articles
	Ecological applications : a publication of the Ecological Society of America	1.87148	Show articles
	Cancer cytopathology		Show articles
	Quality in primary care		Show articles
	Medical image computing and computer-assisted intervention : MICCAI ... International Conference on Medical Image Computing and Computer-Assisted Intervention		Show articles
	F1000Research Open access PubMed Central: after 0 months		Show articles
	Biostatistics (Oxford, England) PubMed Central: immediately	2.40946	Show articles

FIGURE 7.1
Suggested journal names from JANE.

7.3.3 Tips for Formatting

Microsoft Word's spelling and grammar check is also a useful tool to apply to the manuscript before submission, although one should note that it may not recognise various statistical tests and terms such as Kruskal–Wallis test or Prais–Winsten regression. Thus, care is needed when proofreading and checking the manuscript for mistakes. It is highly recommended that one uses a referencing software (such as EndNote, although other software packages are available) for the manuscript as this will help to save time and effort in preparing the manuscript. It is relatively easy to use EndNote, and as with most software, we normally use 20% of the features 80% of the time. EndNote has a cite, while you write function; an interface with Microsoft Word, such that references can be included easily in the manuscript using an additional tab created within Microsoft Word. There are various instructional manuals and tutorials available on how to use EndNote, including the following tutorial video in YouTube: https://www.youtube.com/watch?v=Oskf6sv5Opw (Date accessed: 6 March 2015). The following manual on "Introduction to Endnote X7" from the University of Salford, Manchester, provides a detailed manual, along with useful tips on using EndNote: http://www.salford.ac.uk/library/help/workbooks/endnote.pdf (Date accessed: 6 March 2015). The basic functions within EndNote one should pick up to get started are as follows:

1. Creating an EndNote library
2. Populating the library by inputting articles and books manually or importing from online databases such as PubMed (recommended) (see Figure 7.2)
3. Inserting citations into the manuscript
4. Formatting bibliography according to the style requested by journal

7.4 Preparing a Manuscript

Statistical journals usually have around five or six sections for the main manuscript after the abstract. In the following segment, key points from two articles (Kotz & Cals, 2013; Quinn & Rush, 2009) are summarised, and they provide useful tips and suggestions on how to structure a paper and more importantly how to write the various sections of the manuscript effectively. In particular, clinical and statistical (methodological) examples are provided to exemplify the points.

7.4.1 Abstract

The abstract is often the first thing in a manuscript that a person reads and is used along with the title of the manuscript in the initial filtering process by

Click here to input new article/book

Click here to import article titles from online databases (e.g. PubMed)

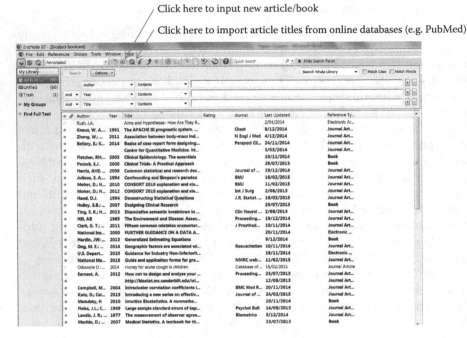

FIGURE 7.2
Inputting journals and articles into EndNote.

journal editors to determine its usefulness and applicability for their readers/peer review. For instance, a researcher looking at evaluating a new statistical method to impute missing data might look for phrases such as 'missing data imputation' in the title/abstract of the manuscript. Abstracts can be both structured and unstructured. The abstract often has the following subsections: background, methods, results and conclusion, with a word limit usually set between 200 and 300 words. Avoid acronyms or abbreviations in the abstract unless they appear more than 3 times. It is also not a good idea to include phrases such as 'the results of the study will be presented' or give lengthy discussions about the literature in the abstract.

7.4.1.1 Tips for the Abstract Section

1. Avoid abbreviations such as SD and statistical symbols α, β and so on. Spell them out in full (e.g. standard deviation, Cronbach's alpha).

2. Don't just present the p-values in the results. Give the effect size (e.g. hazard ratios) as well as the CI.

3. Ensure abstract is revised each time the main text is changed. For instance, planned analysis of a larger sample (additional patients recruited) may provide a new estimate of the effect size, and this

needs to be consistently reported in both the abstract and the main text of the manuscript.

4. Focus on and present results that answer the main research questions. For instance, presenting results from a sensitivity analysis examining the various assumptions without presenting the main comparison between two competing models would be akin to putting the cart before the horse!

7.4.2 Introduction

The introduction is the first part of the manuscript, and it provides readers with information, including what is known about the topic, the current gap in literature or motivating problem as well as the research aims and hypothesis of the study. This is typically written over three to four paragraphs. The first paragraph may perhaps state the importance of the problem. For example, 'Missing data are prevalent in hospital-based administrative records, when they are used for research purposes. Current missing data imputation techniques have not addressed the application to this specific area adequately'. The second paragraph can then introduce the new proposed method or model that aims to address this gap. The final paragraph will state the aims and hypotheses, preferably with a certain level of precision. For example, 'We propose to mathematically derive a new model to estimate missing patient-level clinical variables such as Body Mass Index (BMI) using existing demographic variables and compare its performance in terms of accuracy (evaluated via mean squared error) with existing methods through simulation studies and then applying it to a real-life hospital dataset'.

7.4.2.1 Tips for the Introduction Section

1. The introduction is not a literature review. It is a statement of the nature of the problem and provides a brief overview as to its importance or relevance. It should conclude with the key questions/ hypotheses to be addressed.

2. Do not present any results from this study here, although it is fine to summarise results from other key studies. Similarly, do not discuss the details on methods, design and sources of data in the introduction section.

3. Leave the comparison with other studies to the discussion section of the manuscript. For example, the use of an interrupted time series analysis to evaluate an intervention at a hospital may indicate residual temporal autocorrelations, which may be different to what has been reported previously, and this should be explained in the discussion section, with possible reasons provided to explain the differences.

4. Do not provide too many references. For instance, it is not necessary to summarise the entire literature on sample size calculations for RCTs but perhaps restrict the discussion to more recent established methods. Some people have a mistaken notion that including a large number of references provides greater evidence to substantiate a point, when rather it is the careful selection of the most recent valid argument in literature and discussing that point in the proposal that is important. For instance, the field of missing data imputation has come a long way, and rather than trying to summarise all the available methods out there (including mean imputation or last-observation carried forward), it may be prudent to start the discussion with the most recent established methods (e.g. multiple imputation).

5. Use present tense for established facts (e.g. sample size calculation *is* a crucial component in the design of an RCT) and past tense or present perfect for those that you do not consider as established (e.g. several methods *have been* recently proposed to address the cost-effectiveness of adaptive sample size calculation designs).

7.4.3 Methods

The methods section forms a substantial part of the statistical manuscript, and it usually includes information on the statistical model(s), notation and model formulation, estimation procedure, simulation analysis (if any), data description (including for the simulation and clinical applications) and methods of model assessment. This information needs to be presented in sufficient details so as to enable the results to be reproducible. When established statistical models are presented, they should be referenced. Subsections are commonly used, and this structure varies according to the study and journal's submission requirement. For instance, in a study published in *Biostatistics* journal (Long & Johnson, 2015) that looked at the mechanism of missing at random and variable selection methods which can be combined with imputation, the authors proposed a general resampling approach that combines bootstrap imputation and stability selection. In the paper, they organised the methods into two main subsections (methodology and simulation studies). The 'methodology' subsection was further divided into sections that included bootstrap imputation and an outline of operating characteristics. The 'simulation' subsection contained the details of the parameters and setup of simulation studies as well as the results. Finally, the model was applied to two real-life clinical datasets under a section called 'data examples'.

This is different from the methods section usually found in medical journals, which includes study design, settings and subjects, data collection, outcome measures, interventions, randomisation (if applicable), data analysis and ethical approval (as explained in Section 7.6). Mathematical derivations and results from sensitivity analysis (if any) are usually presented in the appendix section of the manuscript. Some authors present the computer codes that were used in their analysis (e.g. Stata ado codes, WinBUGS, R codes, etc.) in the appendix section as well. Furthermore, some journals make provisions for authors to upload the software codes and even a portion of de-identified data to an file transfer protocol (FTP) website, which is then freely made available over the Internet for other researchers to access.

7.4.4 Results

The results section should mirror the methods section, and for each method, there should be a corresponding result. They should relate to the questions or hypothesis stated at the end of the introduction section of the manuscript. Tables and figures (around five to six) are usually used to supplement the text. The results section for a statistical journal will appear different from a medical journal. Usually, the results from the simulation studies are first reported. This is followed by the application of the new model/method to a clinical dataset (if available). In general, the sections can be separated according to the aims and hypothesis stated earlier in the paper. Results from the simulation studies are best summarised in a table listing the various parameter estimates (e.g. coefficients, standard errors, etc.), which were simulated in the experiment under the different scenarios, along with measures of model fit or evaluation (e.g. mean squared error, bias, median model error, etc.).

7.4.4.1 Tips for the Results Section

1. Use past tense to write the results section.
2. Spell out a number that begins a sentence (e.g. Eighty percent rather than 80%).
3. Do not interpret results to avoid bias. Just report the findings. Interpretation is usually done in the discussion section. For example, with 20% prevalence of diabetes in the random sample, do not discuss that this could be a public health issue in the results section.
4. Use tables and figures to present results but avoid repetitively discussing the results in the text.
5. Provide specific and complete titles and legends for tables and figures. Sometimes, graphs are combined, in which case it is prudent to provide separate titles for the different graphs displayed.
6. Ensure that categories for missing observations are included in the table. This will help readers to get a better idea about the level and nature of missing data.
7. If there are large numbers of tables or figures (generally more than four), consider putting them in the appendix if this is allowed by the journals.
8. Numbers should only be rounded in the final presentation, not during calculations.

9. Make a distinction between statistical significance and clinical significance, and quote the p-values along with CI for the former.

10. When reporting difference in proportions, make a clear distinction between absolute and relative differences in proportion as the difference between the two methods can be substantial!

11. Take care of subscripts and superscripts in the formula and statistical notations.

12. Make sure notations are consistently used across the entire manuscript.

13. Streamline the flow of your writing. Make sure the data from the various parts of the experiments are presented in order preferably under suitable subsection headers.

14. Finally, make sure you present the relevant statistics to answer the research question you sought out to address in the first place.

Example 8

Consider the reporting of the primary outcome measure of improvement in the Dynamic Gait Index (DGI) score among stroke patients randomised to a new physiotherapy intervention (Intervention) versus standard treatment (Control). DGI scores range from 0 (severe impairment) to 30 (normal).

Result. The mean DGI score among the intervention group was significantly lower than the control group (23 vs. 27, respectively), $p = .003$. The above example of writing the results leaves it to the reader to work out the difference and also does not provide measures of variability in the parameter estimates. It is also not clear at which time points the measurements were made.

Revised result. The mean DGI score among the intervention group was 4 units lower than the control group when measured on discharge from hospital (23 [95% CI: 18–25] vs. 27 [95% CI: 26–29], respectively), and this result was statistically significant with $p = .003$.

7.4.5 Discussion and Conclusion

In sum, the discussion section is where one makes use of the statistical results in Section 7.4.4 to address the research question. It is a good idea to summarise the results of the study according to primary and secondary aims in the order they were described in Section 7.4.4. Following that (typically in the second paragraph), the results should be compared against other related studies. If differences were found (e.g. model did not provide an accurate forecast of disease at the small area level), then attempts should be made to explain these differences (e.g. dissimilarity in population profile, geographical setting, setup of simulation studies and the parameters involved). Next, the strengths and limitations of this study should be highlighted. For instance, dependencies between observations may not have been adequately accounted for within the current model or adequate allowances not made for missing observations in the dataset. Strengths of the study such as extensive simulation studies covering a wide range of clinical scenarios and application of the model to a number of different clinical datasets should be emphasised. Sometimes, one may counterbalance the limitation with strength. For example, if a major limitation of the study in terms of excessive missing data has been addressed with a missing data imputation technique in the same study, this should be highlighted.

Finally, the implication of the study results is discussed (e.g. what are the novel findings/contributions to the field, how does the research address gaps in literature, etc.). For example, a new disease forecasting model may have been developed and compared against the existing models and found to be better in terms of accuracy. This may then pave the way for better allocation of health care resources and staffing. The generalisability of the study results should also be described in this section.

7.4.5.1 Tips for the 'Discussion and Conclusion' Section

1. Do not overstate the findings. For instance, if the study demonstrated that the intervention in hospital resulted in a reduction of LOS, one should not necessarily equate that as being cost-effective.

2. Make sure there is a logical link between the hypothesis and the discussion. For example, if a new drug did not demonstrate efficacy even though the study was adequately powered, it would be obviously incorrect to say that it was. Even subtle phrases such as 'almost significant' or 'trend towards significance' should be avoided.

3. Do not present new data in the discussion section.

4. For the conclusion section, do not suggest that further research is needed without being specific about the current limitation of the study that needs to be addressed in the new study.

7.5 Creating a Draft

Statisticians are known to be good with numbers, but it may be a struggle for some to be good with both numbers and writing. This section provides a guide to getting started on writing and creating a draft manuscript. Depending on the personal preference, one may wish to draft a manuscript using the 'quick and fast' method or the 'slow and steady' mode of writing. The former method involves typing the text as fast as possible and not giving much thought to the grammar, style or format of the sections. If one is stuck in the writing, then the idea is to go on to the next section or perform other less strenuous task such as conducting spellchecks and fixing the references. In addition, comments are inserted on points to expound on later. Subsequently, the various sections in the manuscript are revisited and revised. Often several revisions are required to get a decent draft of the manuscript.

However, the 'slow and steady' method involves writing each sentence in each section slowly and clearly and making sure each sentence follows the rest of the paragraph appropriately in terms of style and content. Obviously, such writing style would be slower, but fewer revisions may be needed. However, when major changes are needed to delete or add sections, it may be more difficult to go back and rewrite the earlier sections. Once the draft of the manuscript is ready, it is a good idea to send the paper around to the rest of the collaborators (if any) so that they may add their comments and suggestions on the writing, preferably under track changes. It is also a good idea to use a referencing software such as EndNote so that when paragraphs and sections are moved across the manuscripts by collaborators, any references linked to the sentences are also shifted accordingly, without the painstaking need to individually re-assign them to the referenced text.

7.6 Writing for a Publication – Collaborative Publication

Depending on the nature of the biostatistician's job scope, he/she may be involved in collaborative research with clinicians, which can end up in publications. Often, the publications are for peer-reviewed medical journals found under that particular medical speciality or domain. For example, a pulmonary medicine clinician may wish to publish his/her findings in *Chest* or *Respirology* journal. An emergency medicine physician may wish to publish in *Annals of Emergency Medicine*. Some journals such as the *New England Journal of Medicine* or the *Journal of the American Medical Association* are targeted at more general medical topics. The process of manuscript creation and submission is generally similar to statistical journals in that the journals often have a set of 'instructions for authors' and a peer-review submission process. The main difference between the two types of publications (i.e. statistical vs. non-statistical) is that obviously the statistical journal is targeted at statisticians and those who are mathematically inclined, and hence, it is more technical and methodological in nature compared to the medical journal. Another key difference is that the collaborating biostatistician provides substantial input in the methods and results section, whereas the PI takes on overall responsibility for writing the manuscript for a collaborative project compared to the biostatistician taking on overall leadership for the methodological statistical paper. Section 7.6.1 provides useful tips when writing for a collaborative publication, which complements what has been discussed earlier.

7.6.1 Additional Tips When Writing for a Collaborative Publication

1. Depending on the study design, make use of the guidelines discussed previously (e.g. CONSORT, STROBE) to help the PI write up the methods section of the manuscript. For example, if it is an RCT, describe the randomisation sequence, allocation concealment method, implementation of randomisation and also the blinding process in detail (if applicable) according to the CONSORT statement.

2. Make sure that for each aim and hypothesis listed, there is a corresponding statistical method described, as well as the results presented and discussed.

Example 9

Aim. To examine whether the presence of diabetes mellitus was associated with a higher hospital 30-day mortality rate, independent of age and disease severity among hospitalised systemic lupus erythematosus (SLE) patients.

Hypothesis. Among patients hospitalised with SLE in Austin hospital in Melbourne from 2010 to 2011, we hypothesise that those with diabetes on admission would have an odds ratio (OR) of in-hospital 30-day mortality of 1.5 compared to those without diabetes, after adjusting for age and disease severity.

Statistical methods. Because the main outcome of 30-day mortality is dichotomous, we propose to use the binary logistic regression model to examine whether SLE patients with diabetes will have a higher odds of death compared to those without diabetes. We will include the age and SLE disease activity index measured at admission as covariates in the model. Data analysis will be performed in Stata V13.0 (StataCorp) and the level of significance set at 5%.

Results and discussion. The OR of mortality among SLE patients with diabetes is 2.0 (95% CI: 1.7–2.7) compared with those without diabetes ($p < .001$). The odds of mortality among diabetes patients is 2 times greater than the odds of mortality among those without diabetes, and this was found to be statistically significant, independent of age and disease severity.

1. Make sure that you have included a paragraph on sample size calculation, or in the event of non-significant main result, a power calculation.
2. Provide sufficient details on the statistical analysis to ensure that it is complete and reproducible. For instance, if a step-wise regression modelling was performed, state the probability of variable entry and removal. If a Bayesian analysis was done, explain which priors were used and provide justification and sensitivity analysis, where appropriate.
3. Use tables and figures when presenting data, as they are the most efficient way to bring the information across to the readers. Usually, three to four tables and one to two figures should be sufficient. The first table is usually the demographics, and

Table 1. Univariate factors associated with time from admission till mortality ←Table title. Include *n*
among Chronic Obstructive Pulmonary Disease (COPD) patients (*n* = 1417)

Factors	*n* (%)	HR	95% CI		*p*-value
Age in years: ← Mean (SD)	80.4 (11.4)*	1.62	1.24	1.95	< 0.001$^\sigma$
Male	832 (58.7%)	0.96	0.84	1.08	0.223
Smoker	404 (32.8%)	1.64	1.43	1.89	< 0.001$^\sigma$
Smoking- pack years: Mean (SD)	42.3 (22.4)*	1.08	1.03	1.14	< 0.001$^\sigma$
Comorbid on admission					
Diabetes mellitus	195 (13.8%)	0.93	0.78	1.11	0.234
Renal insufficiency	24 (1.7%)	0.72	0.43	1.18	0.188
Myocardial infarction	244 (17.2%)	0.83	0.70	0.97	0.019$^\sigma$
Respiratory condition	206 (14.5%)	0.74	0.62	0.88	< 0.001$^\sigma$
Metropolitan hospital	1215 (85.7%)	Reference			
Regional hospital	202 (14.3%)	1.11	0.92	1.34	0.264
Private hospital	428 (30.2%)	Reference			
Public hospital	989 (69.8%)	0.57	0.50	0.65	0.013$^\sigma$

← Column headers

Useful to specify units in the title

← Content of table

Note: HR denotes Hazard Ratio
* *Mean (standard deviation) presented*
instead σ *Significant at the 5% level*

Footnote is useful to
describe acronyms
and other unusual
information in table

FIGURE 7.3
Tips on creating a table appropriate for publication.

for a multi-variate analysis, a univariate table is presented before any table showing the multi-variate results. Figures 7.3 and 7.4 provide examples of how to present a table and a figure in a manuscript as well as the important points to note.

4. Look at a previously published article in the journal that the PI intends to submit the paper to, so as to get an idea of the style of writing. In addition, it has also been suggested that it may be easier to write the results section first, followed by the methods, introduction, discussion and summary in that order (Pocock, 2000).

7.7 Resources

1. Grant writers' seminars and workshops (http://www.grantcentral .com/. Date accessed: 19 June 2015). It provides workshops, seminars and consultations on grantsmanship, including providing a writer's workbook with chapters such as 'how to develop a novel, compelling

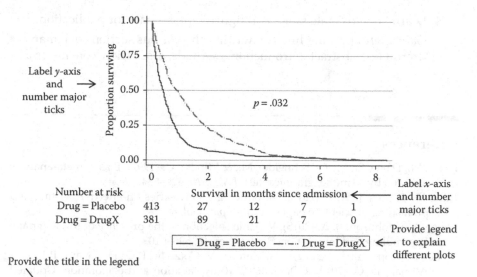

Figure 1. Kaplan–Meier curve showing survival since hospital admission among patients with pancreatic cancer admitted to a tertiary hospital in Melbourne from 2010 to 2011 (*n* = 794).

Note: *p*-value is from log-rank test ← Footnotes are useful to provide more details in the graph, for example, name of statistical test

FIGURE 7.4
Tips on creating a figure appropriate for publication.

idea', 'development of the preliminary studies/previous experience section of your application' and 'the budget and budget justification section of the application'.

2. The following video on 'How to write a great research paper – Seven simple suggestions' provides useful general suggestions on writing a manuscript: https://www.youtube.com/watch?v=g3dkRsTqdDA. Date accessed: 6 March 2015.

Key Learning Points

1. Learn how to write up the various components in a grant application.

2. Acquire useful points on how to write an aim, convert it into a hypothesis and link it in an appropriate statistical method.

3. Understand how the various tools such as the CONSORT and STROBE statements can be used as templates in a grant proposal write-up.

4. Learn how to create tables and figures appropriate for publication.
5. Gain useful tips and hints on writing the various sections of a manuscript (e.g. abstract, introduction, methods, discussion/conclusion).

References

Egger, M., Davey Smith, G., Schneider, M., & Minder, C. (1997). Bias in meta-analysis detected by a simple, graphical test. *BMJ, 315,* 629–634.

Kotz, D., & Cals, J. W. L. (2013). Introducing a new series on effective writing and publishing of scientific papers. *J Clin Epidemiol, 66.*

Long, Q., & Johnson, B. A. (2015). Variable selection in the presence of missing data: Resampling and imputation. *Biostatistics, 16,* 596–610.

Moher, D., Hopewell, S., Schulz, K. F., Montori, V., Gotzsche, P. C., Devereaux, P. J., … & Altman, D. G. (2010). CONSORT 2010 explanation and elaboration: Updated guidelines for reporting parallel group randomised trials. *BMJ, 340,* c869.

Moher, D., Liberati, A., Tetzlaff, J., Altman, D. G., & The PRISMA Group. (2009). Preferred reporting items for systematic reviews and meta-analyses: The PRISMA statement. *BMJ, 339,* b2535.

National Medical Research Council, Singapore. (2015). *Guide and Application Forms for Grant Applications.* NMRC website. http://www.nmrc.gov.sg/content/nmrc_internet/home/downloads.html. Date accessed: 11 February 2015.

Pocock, S. J. (2000). *Clinical Trials – A Practical Approach.* Chichester, U.K.: John Wiley & Sons.

Quinn, C. T., & Rush, A. J. (2009). Writing and publishing your research findings. *J Investig Med, 57,* 634–639.

von Elm, E., Altman, D. G., Egger, M., Pocock, S. J., Gotzsche, P. C., & Vandenbroucke, J. P. (2008). The Strengthening the Reporting of Observational Studies in Epidemiology (STROBE) statement: Guidelines for reporting observational studies. *Rev Esp Salud Publica, 82,* 251–259.

8

Project Management: Best Practices

8.1 Introduction

The number of collaborative projects each biostatistician is concurrently involved in varies, and usually the number is expected to be greater for those who have been collaborating for a while as well as those who have an existing established network of collaborators. The workload can also be seasonal, peaking during times just before deadlines for grant applications and abstract submissions for conference proceedings. This chapter provides a step-by-step guide to biostatistical project management and outlines procedures and processes a practicing biostatistician can adopt to help improve the efficiency of the collaboration process. A distinctive section of this chapter includes the use of a project file that can be created for every collaborative project. This file is useful for the biostatistician to keep track of the process and outcome of collaborations. Files and folder naming conventions are also described, including steps that can be taken to maintain data security and confidentiality, as well as tips on ensuring consistency and reproducibility in the data analysis. The pitfalls in project management (i.e. 'how not-to') are presented in this chapter. Guidelines on managing multiple collaborations as well as ways to identify and work with mentors to better manage collaborations are also described in this chapter.

8.2 Creation of a Project File

It is a good practice to have a project file for each of the substantial collaborative project that a biostatistician is involved in so as to keep track of the process, outcome and deliverables of the project. As the saying goes, if you cannot measure it, it will be difficult to manage it. Substantial work includes collaborative input that leads to a publication, a major presentation or a grant application, but excludes one-off meetings that require a quick advice or consultation. The project file may contain details such as principal investigator (PI) name, project title with a brief synopsis, antici-pated date of publication/grant submission, interim dates for meetings with PI and other project timelines, location of files such as the dataset, related publications, draft grant proposal and expected date of completion. Figure 8.1 provides one such template of a project file. A project number is essential to identify and keep track of unique projects. There may be more than one project from the same PI. Obtaining the PI's contact details is important, in case clarification is needed for the project (e.g. data queries, communicating unforeseen delay in data analysis, etc.) and I've personally found that many are even willing to share their personal mobile number if only one requests for it!

Project # 001

Project title: Identifying factors associated with excess length of stay among patients hospitalised with head and neck cancer in a tertiary hospital

PI name: Dr John Omega.
PI contact details: Department of Radiology, Melbourne general hospital
Email: john_omega@melbgh.com.au, Tel : 03 99130112

Project initiation date: 10 March 2015
Anticipated end date: 20 December 2015
Outcome: ~~Grant~~/ Publication

Location of files:
    1. **Data:** c:\projects\john_omega\data
    2. **Analysis output/software codes:** c:\projects\john_ omega \analysis
    3. **Documents/publications, etc.:** c:\projects\john_ omega \publications

Timeline:

Task	Mar-Apr	May-Jun	Jul-Aug	Sep-Oct	Nov-Dec
Preparation of data	███				
Initial analysis and review		███	███		
Revised analysis				███	
Manuscript preparation					███

Notes:
- PI requested to account for ward-level effects. KIV random-effect models?
- PI still collecting dataset. Will provide updated data on 15 March 2015
- Anticipate submission to BMC Health Services Research journal
- Arrange to meet PI sometime in early May after initial analysis

FIGURE 8.1
Example of a collaborative project file.

Project initiation date is usually the date of the initial meeting with the PI, although for some well-established relationships, projects can be initiated via e-mails/phone calls, provided the biostatistician has worked with the dataset before and is familiar with it. The anticipated end date refers to the publication submission or grant application date, and this date is useful for the biostatistician to work out the timeline and tasks leading up to the deadline. These dates are not carved in stone and are meant to be a guide. They may get pushed back over time. However, penning the dates down leads to a better management of expectations from both sides of the collaboration. This also provides a useful mechanism for the biostatistician to chase up on the status of the project with the collaborator closer to the anticipated submission date, as usually the responsibility for submitting the publication/grant proposal lies with the PI.

The location of electronic copies of files such as the dataset, Stata or other software computer codes, analysis output, related publications/documents and so on should also be included in the project file. It may be tempting and convenient to put all the files in one folder with the hope that there would not be many files and that the files can be sorted out later. However, as experienced biostatisticians would attest to, one would very quickly be swamped with many files and datasets, and these can be cumbersome to sort retrospectively. Sub-folders can be named after the collaborator, or sometimes their initials. Sometimes, it is possible to work on more than one project with the same collaborator, in which case numbers can be appended to the end of the folder name to indicate the order of projects (e.g. john_omega_2, john_omega_3, etc.). Naming of datasets and the structure of files should be standardised. The naming of sub-folders can vary according to individual preference, but the number of sub-folders should be kept to a minimum (maximum 5), and the structure standardised across all projects to ensure consistency. The consistent naming/location of files allows for consistency and reproducibility, and it also permits his/her colleague to take over and continue working on the project in the event that the biostatistician is no longer able to work on the project (e.g. protracted illness or relocation). Figure 8.2 shows an example of a naming convention for a project file/folder. The project folder is named after the PI (e.g. Dr. John Omega). Personally, I find this makes the folder easier to search for rather than using an index or project identifier number. Alternatively, folders can be named after projects (e.g. 'RCT stroke' or 'Predictors no fall'), but it is still relatively more difficult to match projects with the corresponding PIs, especially if there are many collaborative projects. Within the main folder, one may wish to have separate folders for the original data (for reference and quality control), the derived data (data that has been coded and merged or includes derived variables), documents (e.g. proposal, ethics, relevant papers, etc.), software codes and finally the output from the analysis. Depending on the nature of the study, further sub-folders can be created (e.g. 'Simulation results' sub-folder within 'Output' folder to store extensive results from simulation studies).

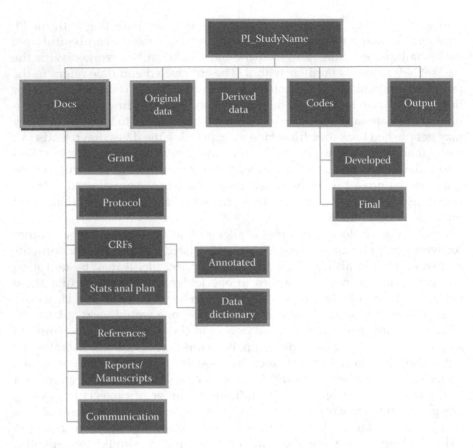

FIGURE 8.2
File naming and folder structure convention for projects.

The timeline or a Gant chart is another useful feature in the project file. It allows the biostatistician to allocate time for the various tasks in order to keep a tab on the overall time management for the project. In the event of unforeseen circumstances that cause a delay in the analysis and reporting of the project, amendments can be done and the final submission date adjusted. The amount of time spent on data analysis varies according to the complexity of the project, and the turnaround time would depend on this as well as other factors such as the current workload of the biostatistician and the time he/she is spending on other projects.

When planning timelines for projects, it is prudent to allow time to make sure the dataset is in a state that is 'ready' for analysis. Some PIs may be mistaken that their data is complete, coded and formatted in a way that makes it conducive for analysis, when in reality it is not. This then causes unnecessary delays in the timeline for the analysis of the data. It is also sensible to allow sufficient time for re-analysis of the data and regular review of the data/analysis with the PI. Finally, the project file should also include special notes peculiar to that specific project. For instance, the dataset might be riddled with missing observations, and the PI may have requested for the possible use of missing data imputation techniques, or the dataset might be a clinical disease registry with the dictionary available over a particular website, in which case it would be useful to note down the uniform resource locator (URL). The notes section can also be used to write out immediate action tasks such as setting up a meeting or presentation. It would also be beneficial for the statistician to discuss with the PI about being included as a co-author or collaborator when the publication or grant is being submitted, and note down perhaps the order of authorship in the document, so as to avoid disputes on authorship later on.

The project file allows the biostatistician to manage individual collaborative projects effectively. When examined at a cross-sectional point of time, data on the total number of active projects the biostatistician is currently working on can be collated, and this information can be useful at the departmental level for staffing, resource allocation and workload projection purposes. However, on a broader level, it may be beneficial to have a separate timesheet (from the project file) that allows the biostatistician to examine how much of time is spent on the various projects. The timesheet can be completed at the end of the week, and perhaps submitted to the department manager on a monthly/quarterly basis or whenever reporting needs to be done for the department. At the Centre for Quantitative Medicine at Duke-NUS Graduate Medical School, Singapore, such timesheets have been used to determine overall time faculty and members spent on projects at the various institutions (e.g. Singapore General Hospital, National Neuroscience Institute, etc.) as well as other university teaching programs (e.g. third-year medical students, research development seminars, etc.). Such information is then used to distribute work load across programs and staff. In addition, when substantial effort is recorded consistently from a particular institution (e.g. at least 0.5 full-time equivalent of a statistician's time on projects), information from the timesheet is used to justify and negotiate for a regular payment from the institution (i.e. through secondments or service contracts).

An example of a timesheet is shown in Figure 8.3. The timesheets can be used as compelling evidence to present to higher management for hiring of new staff, especially when the number of collaborative projects increases and remains high. It should be noted that the timesheets should not be used to evaluate an individual staff's performance or make productivity comparisons across staff as this is not the primary purpose of the tool. Such use may demotivate staff and cause inaccurate entries for time spent on project, thus invalidating the actual use of the timesheets. The timesheets may also not have all the necessary information on the usual academic key performance indicators such as publications, grants and teaching activities in them. Time spent on projects may also not be the best indicator of productivity or efficiency. The purpose of the timesheet should be communicated to all staff so that the actual hours spent on project are clocked in by staff regularly.

Time Tracking Summary for month: Nov 2015

No	Name (% Paid)		General	Medical students	MPH program	PhD mentorship	Independent Research	Mentoring	Hospital A	Hospital B	Hospital C	Hospital D	Software classes	Others	Total
1	Professor Mean	Hours (%)	6.0	38.0	6.0	5.0	–	–	–	6.0	–	40.5	–	23.0	124.5
		Actual	4.82%	30.52%	5%	4%	0%	0%	0%	5%	0%	33%	0%	18%	
	100	Allocated	5%	55%	10%			15%		5%			10%	–	
2	A/Prof. Keen	Hours (%)	3.0	–	2.0	–	25.0	–	–	–	13.0	2.0	–	–	45
		Actual	6.67%	0.00%	4%	0%	56%	0%	0%	0%	29%	4%	–	0%	
	60	Allocated	5%		20%		20%				15%			–	
3	Dr Holdwater	Hours (%)	81.0	4.0	4.5	–	–	–	–	–	–	–	–	–	89.5
		Actual	90.50%	4.47%	5%	0%	0%	0%	0%	0%	0%	0%	0%	0%	
	100	Allocated	5%	20%	25%						50%			0%	
4	A/Prof. Subra	Hours (%)	81.25	0	4	0	0	0	0	0	0	0	0	56	141.25
		Actual	57.52%	0.00%	3%	0%	0%	0%	0%	0%	0%	0%	0%	40%	
	70	Allocated	10.00%		10.00%									50.00%	
5	Dr Gan	Hours (%)	–	–	2.0	–	4.0	–	3.0	–	–	–	–	6.0	15
		Actual	0.00%	0.00%	13%	0%	27%	0%	20%	0%	0%	0%	0%	40%	
	15	Allocated					15%		15%					40%	
6	Dr Lisa	Hours (%)	27.0	–	–	8.0	145.0	–	–	–	–	–	–	9.0	189
		Actual	14.29%	0.00%	0%	4%	77%	0%	0%	0%	0%	0%	0%	5%	
	100	Allocated	5%				85%						10%		
7	Dr Wan Ping	Hours (%)	47.0	15.0	21.0	–	19.0	–	–	–	7.0	–	–	–	109
		Actual	43.12%	13.76%	19%	0%	17%	0%	0%	0%	6%	0%	0%	0%	
	100	Allocated	20%	10%	10%		10%				50%			0%	

FIGURE 8.3

Example of timesheet with time spent by projects.

8.3 Database Security and Confidentiality

When working with databases and projects, it is essential to maintain the confidentiality and privacy of patients' information as well as take steps to maintain database security. In fact, this is often a requirement from most university's and hospital's institutional review boards (IRBs) before approval is given to undertake a research study. For example, in the University of California, San Diego, California, under section 3.6 of the standard operating procedure of the Human Research Protections Program, it is clearly stated that appropriate plans to protect the confidentiality of research data, an informed consent form that discloses the risks to privacy and confidentiality and physical safeguards for research data are clearly stated in the research project (https://irb.ucsd.edu/3.6.pdf. Date accessed: 10 March 2015). However, there needs to be a balance between the IRB's mandate to maintain confidentiality of patients' health records and the ease of undertaking clinical research by to solve real-life clinical problems. For the biostatistician who works on a large number of patient datasets in the course of his/her collaborations, the following steps provided in Sections 8.3.1 and 8.3.2 to maintain database confidentiality and security, respectively, are recommended:

8.3.1 Database Confidentiality

1. Always keep a de-identified dataset for the analysis. Identifiers include the following:
   a. Name
   b. Address
   c. Date of birth/age
   d. Telephone number, e-mail address, etc.
   e. Fingerprints, facial images, etc.

   A full list of the 18 recommended patient identifiers is given in the Health Insurance Portability and Accountability Act (HIPAA) guidelines here: http://www.hhs.gov/ocr/privacy/hipaa/understanding/coveredentities/De-identification/guidance.html#standard (Date accessed: 10 March 2015).

2. If identifiers are required for research purposes (e.g. spatial analysis of lung cancer cases based on residential address or merging datasets from various sources using probabilistic linkage methods), keep the identifiers in a separate location from the rest of the dataset.

3. Consider aggregating data, wherever possible, before extracting data for analysis. For example, individual address data can be aggregated at some spatial level (e.g. postal code, statistical local area) or disease counts can be summed up monthly for a time series analysis.

4. If a collaborator accidentally sends you a dataset with patient identifiers, immediately delete the file and request for a de-identified dataset to be sent instead. Remind the collaborator to send de-identified datasets in future.

5. Avoid sharing datasets with other research personnel (e.g. research assistants or clinical coders) unless they have been listed as collaborators in the ethics application.

6. Do not print out datasheets with patient identifiers and leave them unattended in the office.

7. Ensure that you do not send sensitive data to your intended recipient (i.e. collaborator) through an intermediary (e.g. research assistant) without taking necessary steps to password-protect the file.

8.3.2 Database Security

1. Consider using file encryption software such as WinZip or even password-protect features designed for spreadsheets such as Microsoft Excel, which locks data by cipher.

2. Write-protect the data and program files to avoid accidental deletion or the possibility of someone accidentally overwriting the files. To do so, right-mouse click on the file, select 'Security' and then make changes to the read/write status of the file accordingly.

3. Keep sensitive datasets on stand-alone PCs without any Internet connection to ensure that hackers cannot access the data.

4. Provide different log-in IDs (access controls) for separate personnel of the study team (e.g. PI, biostatistician, research assistant) so as to enable an audit trail on who has accessed the dataset.

5. Always lock your PC/laptop when you leave your workstation, especially if your office cannot be locked.

6. Minimise sending databases over e-mails and File Transfer Protocols, wherever possible.

7. Do not remove software codes and/or analysis datasets when leaving employment, unless there is prior approval from the PI.

8. Undertake proper handover of projects and datasets to colleagues/department before leaving employment.

9. Keep accurate records of data and, if possible, inventorise the projects regularly (e.g. use dated files, versions).

8.4 Standard Operating Procedures

The creation of standard operating procedures (SOPs) may initially appear as an unnecessary administrative task, but properly crafted SOPs can actually aid biostatisticians in managing projects. For instance, there could be a SOP specifically designed to guide collaborators on how they engage in a biostatistical collaboration for a grant proposal. Figure 8.4 lists down the procedures, timelines, deliverables and reference documents related to an exemplifier SOP for biostatistical input in a grant proposal. The timeline provides a useful reference for both the biostatistician and the collaborators to plan individual tasks, and also serves as a reminder for the grant planning some months in advance. Four months (D-4) from the final submission deadline of the grant is a reasonable timeline. Much of the details in the grant proposal (e.g. logistics, budget, etc.) would depend on the study design and the final sample size, so the PI would need ample time to work out the rest of the details after finalising the study design and methodology with the biostatistician, and this point must be emphasised to the collaborator.

Allocating timelines for an individual project would also allow the biostatistician to know how much time he has to work on other grant proposals and can improve the efficiency by undertaking shared tasks simultaneously (e.g. updating common information in grant applications such as PI track record, shared programming codes and undertaking power calculations). The reference items in the SOP aim to improve the quality of the first draft of the grant proposal before the biostatistician has a look at it. Obviously, the quality of the initial draft proposal would depend on the research experience of the PI, but these documents serve to reduce avoidable delays in discussions around issues such as 'writing down the research question and aim' in a way that is specific enough to translate into a workable hypothesis. The suggested document on 'sample size ingredients' also allows collaborators to search the literature for relevant information such as effect size and pilot data, which can then be used in the sample size calculation process when meeting up with the biostatistician. This saves valuable time and ensures a more fruitful collaboration. A similar SOP can also be created for biostatistical input leading up to a publication. In such a SOP, guidelines can be provided as to the preparation of the dataset (e.g. required format and structure of dataset), the scope of work included for the biostatistician (i.e. data analysis, writing up of manuscript) and perhaps even work exclusions (e.g. creation of posters, data entry, data management, etc.). Turnaround times for data analysis as well as deliverable items can be listed in the SOP. If the department is large enough, it may also be possible to have different biostatisticians assigned to various medicine specialities (e.g. respiratory medicine, psychiatry, etc.) and make this information displayed in the SOP.

Biostatistical input for a grant proposal

Statement of Intent

The Biostatistics Unit has been mandated to improve the success rate of grant proposals submitted by Principal Investigators (PIs) from Melbourne General Hospital. This document aims to set up biostatistical consultancy procedures in place to achieve this goal.

Procedures

PIs will provide all relevant input and adhere to the timelines stipulated below, in order to efficiently engage biostatisticians prior to submitting their grant proposal. They will read and be familiar with all relevant documents highlighted below. In return, they will receive high-quality biostatistical input described under deliverables.

Timelines

Time	D-4 (4 months prior to grant submission deadline)	D-3	D-2	D-1
Task	• PI reads essential documents (see below) and provides requested information • PI submits draft proposal to biostatistician • Initial meeting with biostatistician (optional)	• PI meets with biostatistician and goes through initial input • Outstanding issues clarified	• Biostatistician provides revised draft	• Biostatistician provides final draft

Deliverables

Biostatistician will provide timely quality input in the grant proposal, particularly in the design, specification of the hypothesis, analysis methods and sample size, as evidenced by less than 2 queries from peer-reviewers in these areas.

Reference documents

1. How to write a successful grant proposal
2. Ingredients of a good research question
3. Aims and hypothesis-How are they linked
4. Guidelines on study design and analysis
5. Inputs required for sample size calculations

Author: Department manager, Biostatistics Unit, Melbourne General Hospital	
Date issued: 25 Jan 2015	Next review: 25 Jan 2016
Date revised: NA	File location: C/Biostats/SOP/grant_sop.doc

FIGURE 8.4
Example of a SOP for a grant proposal collaboration.

It would also be helpful to have a separate SOP on the storing of electronic datasets from collaborators. Guidelines on naming files and folders along with steps to ensure data security and confidentiality can be included in the SOP.

These SOPs should be created preferably with the input of all the biostatisticians in the department. They should then be made widely available to all collaborators (e.g. on a shared folder on the intranet) and regularly disseminated (e.g. through flyers and emails) so that awareness among collaborators is increased. By understanding the steps and procedures set in place for a biostatistical collaboration in a SOP, collaborators would be discouraged in approaching the biostatistician at the last minute. This would be particularly useful because the biostatistician is usually going to be inundated with many requests for collaboration closer to the grant submission deadline, and it should also be noted that last-minute work on proposals will likely lead to a poorly developed proposal that is unlikely to get funded. The SOPs can also be a vehicle to ensure that the biostatistician gets recognised in the proposal or publication by being included as a co-investigator/collaborator or co-author. SOPs also provide information on the job scope and workflow involved for a biostatistician in the academic setting and is an excellent source of orientation guide and on-job guide for new biostatisticians.

8.5 Ensuring Consistency and Reproducibility in the Results

A biostatistician often works on multiple collaborative projects and datasets and sometimes with multiple versions of the same dataset. Hence, it is important that good principles of project management are followed. In particular, if a project involves data analysis, it is important to ensure consistency and reproducibility in the results. This will help to increase efficiency in project management and also enable quality control audits to be undertaken and allow for easy handover of projects to colleagues. Through the process, the research project's (and institution's) credibility is enhanced. Some journals may require data to be submitted as part of the manuscript submission process or for the author to provide the data whenever requested. For instance, under the instructions for authors, the *Journal of the American Medical Association* (JAMA) states that 'If requested, authors should be prepared to provide the data and must cooperate fully in obtaining and providing the data on which the manuscript is based for examination by the editors or their assignees' (http://jama.jamanetwork.com/public/instructionsForAuthors.aspx. Date accessed: 18 June 2015). This section provides some points on how one can ensure consistency and reproducibility in the data management, analysis and reporting of a research project. Examples of software codes will be provided.

1. *Keep all relevant datasets in the same folder.* Data for a particular project may need to be extracted from multiple datasets or sheets (e.g. different sheets for the various years of data collected or patient demographics and laboratory results originating from different sources, which then need to be merged using a common patient identifier). It would be preferable to keep all the working files in the same folder so that it is not necessary to specify the location of the files for each of the command typed in reference to the files. The current directory that Stata is working on should be changed to the project directory (e.g. cd "C:\ Dr_Davids\data"). With all the source files located in one folder, one is also less likely to have broken commands when copying project folders to another media (e.g. thumb drive or a back-up drive). It is also easier to locate the files in future. Keeping all data in a single folder also makes it easier to locate files and back them up regularly. In addition, one should try and preserve the original datasets as much as possible, and create and work on derived datasets instead (e.g. in Stata, one can write 'do' files to merge multiple datasets (type 'help merge').

   Similarly, one should document any changes made to the dataset (e.g. by including comments in the Stata 'do' command file and printing and filing e-mail instructions) and make sure that the most recently updated dataset is used in the analysis. Sometimes, it is possible that the data is inadvertently changed (e.g. data amended accidentally or a variable dropped and the original file then overwritten). As an audit trail, one can take note of when the dataset was last saved, but Stata also has a useful feature to help detect whether the file has been changed. By typing 'datasignature', Stata will return a unique set of numbers associated with the dataset that will change even when small changes are made to the dataset. To find out what exactly has changed between two versions of a dataset, one can use the cf command (e.g. type use 'version1.dta', and then 'cf _ all using version2.dta'). Stata will indicate which variables have been added/removed as well as the specific changes made to observed values within the variables.

2. *Use syntax commands and keep a copy of the commands.* Using syntax commands helps ensure reproducibility in the analysis. In Stata for example, this is achieved by saving the commands as a 'do' file and then running the commands in the 'do' file editor. Figure 8.5 demonstrates how this can be done. In addition to ensuring reproducibility, this feature is also useful in improving the efficiency of the data analysis process, particularly when changes have been made to the dataset (e.g. inclusion of new variables or observations). One should note that if data management functions are written in the 'do' command file (e.g. recoding, creating new variables

Opening a 'do' file editor

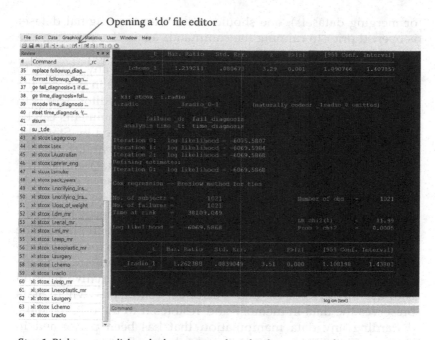

Step 1. Right-mouse click and select commands under the review panel you want to include in the 'do' file. Tip: use the shift and control key to select multiple variables that are consecutive and non-consecutive respectively.

Step 2. Click on 'do' file editor (see arrow above). Paste commands and save 'do' file. To run the commands, click on the 'execute' icon (see arrow below).

Executing a 'do' file

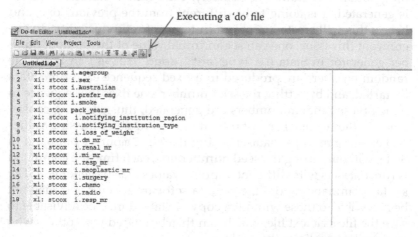

FIGURE 8.5
Creating and running a 'do' file in Stata.

or merging datasets), one should make sure the original dataset is opened prior to running the commands again (i.e. recoding age into age groups, and then running the command again will give incorrect age classifications!). In a similar fashion, any changes to the data (e.g. incorrect data entry) should be made in the 'do' file and not changed manually in the spreadsheet file editor. This is to ensure that such changes are well documented. Comments can be inserted into the 'do' file (these will not be executed by Stata) by typing an asterisk before the text (e.g. * Age for Patient ID Study_07 was changed from 49 to 19 because of incorrect data entry). The cmdlog file can also be generated to review and save previously executed commands for a project (type 'help cmdlog' to get more information). Sometimes, subgroup data analysis needs to be performed (e.g. separate analysis by hospitals or demographics). One option is to use the 'if' command to select the subgroups of interest, but this can be cumbersome, especially if there are multiple commands to execute for each subgroup. An alternative method is to use the 'preserve' and 'restore' commands, where Stata preserves the data in memory and restores it to the original data, discarding any data manipulation that has been performed in between. This will be particularly useful when there is an extensive range of data management functions that need to be performed for each subgroup analysis.

Sometimes, it is necessary to generate random numbers (e.g. 'generate random = uniform()') for random sampling or running simulation studies. Obviously, each time a random number is generated, it is going to be different from the previous one, and hence, this will create a problem in terms of reproducibility. To circumvent this issue, one can set an initial value of the random number generator in Stata (i.e. type 'set seed 12596'). In Stata, the random numbers are produced in a fixed sequence each time Stata is started, and by setting the seed number, one fixes the sequence in which these random numbers are generated, thus ensuring consistency in the results. For instance, if one is imputing missing values and creating multiple datasets using the 'mi impute' command in Stata, without setting the seed number first, each time the command is run, datasets with different imputed values may be generated. The syntax command or 'do' file is Stata software specific and may not be accessible to those without a copy of Stata. It may be advisable to save the file as a text file, which can then be shared with others who may wish to replicate the analysis.

3. *Create 'log' files*. One can also save the command and results in a log (text) file, which can easily be disseminated to collaborators and others who may not have Stata software. There are three simple

steps to create a 'log' file within Stata. Firstly, one creates a new log file in the 'do' file editor by typing 'log using analysis.log, replace'. The filename is 'analysis', and the 'replace' option informs Stata to replace the original file whenever the command is rerun. Instead of 'replace', one may also opt for the 'append' option where additional analysis output is added to the earlier analysis that has been performed in a cumulative manner. The second step is to type in all the commands that one wishes to run for the project. The final step is to close the log file by typing 'log close'. In the current folder that Stata is working on, one should be able to find the file 'analysis.log', which can then be opened with Notepad, Microsoft Word or other similar word processing software. When saving the file as a Word document, it may also be useful to date the file (e.g. Analysis 20 April 2015) in order to identify the latest analysis file and differentiate it from earlier copies sent to collaborators.

4. *Avoid manual copying from Stata output.* It is not advisable to manually transfer the output from Stata into a text file or an Excel table, as mistakes may inevitably occur. There are other ways to do this more efficiently and accurately. If the analysis involves multiple regression models (for example), and the regression coefficients along with their confidence intervals and *p*-values need to be extracted into a table (as required by most journals), then the text-to-column function within Excel can be used. Figure 8.6 demonstrates how this can be done. Firstly, the log file is created. Then, open this text file from within Excel (remember to change file type to 'all'). The text import wizard will appear (Figure 8.6). There are two ways in which Excel separates the text file into tables: delimited (where characters such as * or | can be used to separate data into separate tables) or fixed width (vertical lines are drawn through the text file to separate fields). Here, select 'Delimited' and click on 'Next'. In the second step, select 'Space' and click on 'Next' and then 'Finish'. In the resulting Excel sheet, all the data would be put in separate columns as required. After deleting the unnecessary rows and a bit of formatting, publication quality tables can be created (see example Table 8.1 and Figure 8.6). User-written programs such as 'outreg' also exist. Outreg allows for the necessary output from the regression model, such as the regression coefficients and their standard errors, along with asterisks to denote statistical significance to be displayed in a word document. Type 'findit outreg' under the Stata command to find out more about this program and how to install it. Programs such as 'outtable' and 'outtex' are also available to automate the conversion of a Stata matrix into a LaTeX table.

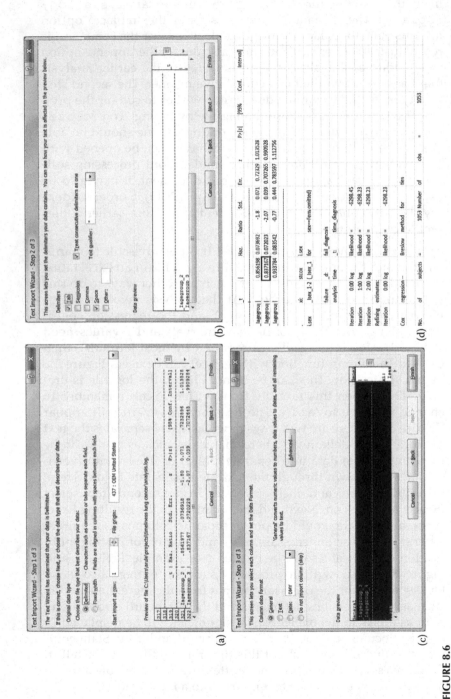

FIGURE 8.6

Transferring Stata output into Excel tables. (a) Step 1 of importing text to Excel spreadsheet. (b) Step 2 of importing text to Excel spreadsheet. (c) Step 3 of importing text to Excel spreadsheet. (d) Step 4 of importing text to Excel spreadsheet.

TABLE 8.1

Factors Associated with Survival Among Prostate Cancer Patients

Factors	HR	95% CI		p-value
Female	Reference			
Male	0.96	0.84	1.08	0.232
First treatment intent				
Non-curative	Reference			
Curative	1.23	1.09	1.40	<0.001
Loss of weight				
No	Reference			
Not stated	0.95	0.67	1.35	0.843
Yes	1.04	0.92	1.18	0.233
Comorbid from medical records				
Diabetes mellitus	0.93	0.78	1.11	0.101
Renal insufficiency	0.60	0.43	1.18	0.294
Myocardial infarction	0.83	0.70	0.97	0.002
Respiratory condition	0.70	0.62	0.88	0.010
Coronary heart disease	0.92	0.79	1.07	0.789

8.6 Managing Multiple Projects

At any time, a biostatistician is usually working on multiple projects. The actual number of projects may vary seasonally, depending on whether there are any impending deadlines for grant submissions, conference proceedings or even calendar dates for medical doctors to pass their medical specialty training (and hence the urgency in publishing). In terms of the actual work involved, this can also vary considerably: ranging from reading up on grant proposals and journal articles, to performing sample size calculations, data preparation and statistical analysis. As the survey in Chapter 11 found, the median number of projects biostatisticians collaborate in a 12-month period is seven.

The key to successful management of multiple projects lies in effective time management, prioritising projects and keeping track of milestones for each project. Effective time management includes spending the least amount of time to achieve a particular goal or outcome (i.e. getting a paper published or grant funding). In terms of collaboration leading up to a publication, this includes drafting the appropriate statistical analysis plan early during the collaboration and making sure that the plan is mutually agreed upon by the PI and other collaborators in the team. It would be a waste of time to jump straight into the analysis and work for, say, 4 days on the analysis, and then for the PI to request for a re-analysis because the primary outcome measure has been changed or to accommodate anomalies in the data that could have been discovered earlier (e.g. missing data requiring imputation techniques). The statistical analysis plan should also be shared with the team early during the collaboration (e.g. via documentation in the project file or communicated through e-mail), and the analysis started only if there is no objection to the agreed plan.

The statistical analysis should also begin after the receipt of the 'final' and 'cleaned' dataset. It would be inefficient to perform the analysis and create tables and other output when, say, 90% of the data has been collected, and then to repeat the entire process once the PI has collected the remaining 10% of the samples. Similarly, it would be frustrating to rerun the analysis after discovering extensive mistakes in the dataset (e.g. data entry or coding errors) that could have been identified and rectified beforehand. The statistician should allocate some time to perform descriptive statistics and undertake queries and quality control checks on the data before analyzing the data. For the initial analyses, the biostatistician can also consider sending the collaborator raw outputs from the statistical software and defer transposing them into proper tables and figures until the collaborator is satisfied with the output as this can be time-consuming.

Time-consuming data management tasks such as recoding of continuous variables into categorical ones, generating new variables and generating and formatting date variables should preferably be performed using statistical software such as Stata, rather than relying on Excel or other data management tools. The extensive programming capability within statistical software such as Stata can also be used to process multiple and repetitive commands. For example, if 100 independent Student's *t*-tests are required to compare parameters between two groups (e.g. males vs. females) in a study (note: this is rarely the case in a well-designed study!), substantial time can be saved by using the loop command in Stata, described below:

```
foreach x in var1 var2 var3-var100 {
ttest `x',by(gender)
}
```

It is helpful to schedule work at different times of the day, depending on the scope and complexity. For instance, mentally stimulating tasks such as writing software codes for data analysis can be performed in the earlier part of the morning, and more mundane tasks such as creating tables and figures can be done towards the end of the day. Writing for manuscripts or grant proposals should be done in blocks of time (e.g. 2 h) to keep the flow of thoughts going, but regular breaks should be scheduled (e.g. every half-hour) to break the monotony. Prioritising projects is another important component of multiple project management. Lining up a long list of projects to work on during the day is not necessarily a good strategy as it may lead to burnout. Urgent projects may provide more stress with their impending deadlines, but they may not necessarily be the important ones.

A hastily put together analysis will seldom see get the paper getting published in a good journal, and a last-minute grant proposal usually leads to a low probability of success of securing funding. Projects may be classified into one of the following four compartments depending on their importance and urgency (Figure 8.7). At the beginning of each month (or appropriate interval), it would be useful for the biostatistician to assign the various projects into this matrix, and then prioritise the projects accordingly. For instance, projects where the biostatistician has been named in a grant proposal or publication and where the deadline for submission is impending can be classified into box A. These are projects with the highest priority. On the other end of the scale, poster presentations may fall into box B or D, depending on the urgency as these have relatively less weightage in terms of academic key performance indicator and should have low priority.

If the number of projects in box B piles up, then some of these can be shifted to box C by negotiating with the PI for a longer turnaround time. Items in box D should be avoided whenever possible. For instance, department SOPs can be used as justifications to politely decline projects where a poster presentation is the only intended outcome of a collaboration. For PIs with multiple projects, the statistician should work on one project at a time and wait for submission of the work before starting on the next project with the same PI. This is to avoid ending up with multiple ongoing projects with no project closure for any of them. Some projects can naturally shift from box C to

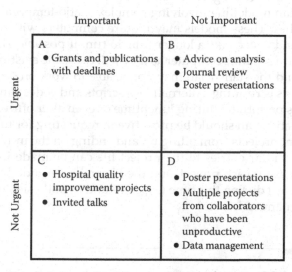

Note: Assignment of projects to quadrants may vary depending on the job-scope of the statistician and aim/mission of the department

FIGURE 8.7
Urgency and importance matrix for collaborative projects.

box A over time. For instance, a hospital quality improvement project on factors associated with re-admission can be turned into a research project and eventual publication and upgraded to box A, giving it a higher priority. Similarly, substantial statistical input into a project can be used to justify authorship and the shift of projects from box B to box A. If a biostatistician finds that most of the projects end up in box D, this should be discussed with the department manager, and his/her job scope as well as the department's goals should be reassessed so as to better manage incoming requests for bio-statistic collaborations. However, if too many projects come under boxes A and B, departmental SOPs on workload distribution should be re-looked and the possibility of recruiting additional staff evaluated.

Another useful tip in effective project management is to avoid doing other people's work. For instance, research assistants are often employed by PIs to collect data, input the data into spreadsheets and even manage these datasets. Although it is important for the biostatistician to have some understanding of the nature of the data by performing descriptive analysis, other tasks such as coding and labelling of variables should be left to the research assistant. Similarly, the biostatistician should not be involved in the actual design of the questionnaire or case report form but rather provide critical input in the choice of questions or variables to be included in such forms. Other tasks such as development of the protocol for the research project or ethics application are usually undertaken by the PI, but the statistician may just be required to provide a brief and specific input in the design and analysis section.

Some projects may need greater computational resources than others, for example, Bayesian modelling involving complex spatio-temporal analysis or simulation studies. These models may require computers with a faster processing speed and hence take a longer time to run. If possible, these models should be run on a separate computer as they will cause the existing computer to slow down and the statistician may not be able to work on other projects, while the analysis is running. Alternatively, scripts and codes can be written, and the analysis performed during lunchtime or even after office hours.

Finally, the statistician should be proactive in requesting for updates. This will help prevent projects from piling up and ending on the in-tray of the PI. For instance, calendar entries to the project file can be made in relation to critical milestones for each project, and the PI can be reminded to provide a status of the project during this time (e.g. aims and hypothesis, comments on analysis, draft manuscript, etc.).

8.7 Obtaining Mentorship

For a biostatistician who is just starting out work in the academic setting, perhaps the most efficient way of learning how to manage projects and relationships with collaborators is to obtain guidance from the more senior or established colleagues in the department or sometimes even from experienced clinicians. When choosing a statistical mentor, there are a number of factors one should consider. At the outset, it may be tempting to select the most senior person in the department or someone who has published widely. However, one should also make sure that the mentor is available and willing to spend enough time with the mentee, especially on urgent and difficult issues relating to a collaborative project. It would help if the mentee and the mentor can be included as co-authors in a publication, and this could be communicated to the mentor to potentially get buy-in.

The mentee should also critically evaluate gaps in his/her knowledge and skills before trying to find a mentor who may be able to help. For example, one may be technically adept in the various biostatistical methods but lack confidence in their communication or project management skills. It would then be better to select a mentor who has extensive experience in collaborating with clinicians and who is willing to accompany the mentee, at least during the initial meetings with the clinician. The mentee should then listen and learn from his/her mentor how he/she communicates and negotiates with the clinician during discussions. The biostatistician can also have more than one mentor, perhaps one who is technically adept and the other good in communicating with clinicians. The mentor may not necessarily be a biostatistician as well. Clinicians who are experienced with working with statisticians can also perform the role of a mentor, providing advice from the clinician's perspective and sometimes even sharing statistical techniques, which are specifically used in their medical fields (e.g. agreement statistics in radiology). Larger institutions may have a policy of assigning mentors to new staff, but for smaller statistical units, sometimes it may be necessary to source for mentors elsewhere.

If the mentee is keen on developing his/her area of interest, then it would be wise to seek a mentor who has done substantial work in the field. For instance, if one would like to work in the area of 'statistical methodology related to vaccine or nutritional research', it might be better to search in statistics or biostatistics departments for potential mentors who are experienced in these fields. Search engines or department websites can be used, and sometimes mentors can be approached at statistical conferences when they present their latest research. It may be possible for mentors to be located in a different university or even a different country. Depending on the complexity of the problem and the availability of the mentor, the biostatistician may sometimes need to relocate or find a way for his/her mentor to host him in his/her institution (e.g. PhD or postdoc positions).

Desirable qualities of a good mentor include the following (Berk et al., 2005; Thabane et al., 2007), and these should be considered when looking for a mentor:

1. Competence
2. Respect
3. Power/influence

4. Experience

5. Political acceptance

6. Honesty

7. Established record of mentoring

8. Confidentiality

After the mentor has agreed to the mentorship, the mentee should be proactive in managing the relationship with the mentor. Mentorship can be time consuming and may strain the relationship if the expectations from both sides are not carefully managed. The mentee should draw up a schedule listing the specific areas, time and input required from the mentor. This schedule should be regularly revised and updated with input from the mentor. For example, fortnightly 1 h meetings can be arranged where the mentee presents challenging statistical problems that he/she needs guidance from the mentor. During the meeting, the mentee needs to explain the specific input he/she would like from the mentor (e.g. suggestions on alternative analysis plans, correction of software programming codes, etc.). It is also essential that the mentee follows up with the actionable items and reports this to the mentor during the next meeting. Nothing can be more annoying to a mentor than to find out that all the advice and input given previously has been ignored. In order for the relationship to work, the mentee also needs to take any criticisms positively. Should there be disagreements with the advice provided by the mentor, these should be carefully brought up and if possible justified with appropriate references (e.g. highlighting journal articles showing the perils of analysing change scores due to the 'regression to the mean' problem).

Sometimes, a mentor–mentee relationship may not work out, and it would then be necessary to end the mentorship arrangement. This could arise from a number of reasons, including the mentor no longer having time to provide mentorship, mentor and mentee not matching up to each other's expectations, mentor being too demanding or aggressive, mentee not following advice or suggestions given by mentor and so on. The mentee should then evaluate what were the reasons for the relationship to fail and then try and find another mentor who can fill the gap. Often, mentorships work well and reach maturity, when the mentee is able to work and function independently. At a later stage, the mentee should also consider providing mentorship to others who may then gain as much as he/she did.

8.8 Poor Project Management Skills (The 'Not's to Avoid)

1. *Not prioritising projects.* Statisticians may sometimes find themselves working on urgent projects that are not important if they do not take stock of the projects regularly.

2. *Not saying no often enough.* Sometimes, statisticians may find it difficult to decline new requests for collaborations even when they are inundated with many projects. As Stuart Pocock (1995) aptly describes in his article, 'The frustration of overcommitment is enhanced by one's inability to decline certain tasks. We are, on the whole, a helpful profession, and hence the art of knowing when to say "no" does not come easy to many of us'. It is fine to tell collaborators that it may not be possible to work on their projects until a couple of weeks/months later or even politely reject projects if there are too many currently on hand. Providing reasons including 'I may not be able to spend enough time on your project as I would like to' or 'There is an important deadline for the other project I'm working on' could help.

3. *Not documenting projects.* It is important for statisticians to organise a project file that states the location of files and folders and also document the statistical analysis plan, data dictionary and any other issues pertaining to the project. This is particularly useful when there is a long time lag since the last time the statistician has worked on the project (e.g. through lengthy publication process or collaborator keeping project in the back-burner for a while), and it makes it easier to pick up where one has left off.

4. *Not having clear communication with collaborator.* Input from the collaborator should be obtained in the early stages of the project and also regularly enough, especially when deliberating on the sample size calculation or statistical analysis plan. These should also be communicated with the collaborator in a language that is understandable. The statistician should also be proactive in contacting the collaborator and obtaining status updates for the project regularly.

5. *Not using statistical software efficiently.* Data analysis is usually not a one-off task. Writing a software command code (e.g. Stata ado file) to perform data management and statistical analysis tasks can make such tasks more efficient. Loop commands and 'by'/'if' statements can be used to perform repetitive steps quickly.

6. *Not utilising SOPs.* SOPs are only useful if they are widely disseminated and used consistently. For instance, SOPs on early engagement of statisticians for collaborative input in a grant submission can be used to discourage those who approach the statisticians at the very last minute.

8.9 A Collaborative Case Study

Project title. Ethnic variation in the correlation between fasting glucose concentration and glycated hemoglobin (HbA1c).

Collaborating clinician. Dr. Dalan Rinkoo, MBBS, MRCP (London), FAMS (endocrinology), FRCP (Edinburgh); senior consultant, Tan Tock Seng Hospital, Singapore.

Output. Following publication. Dalan R, Earnest A, Leow MK. Ethnic variation in the correlation between fasting glucose concentration and glycated hemoglobin (HbA1c). *Endocr Pract.* September–October 2013;19(5):812–17. doi:10.4158/EP12417.OR.

Study summary

The objective of this collaborative project was to examine the relationship between fasting serum glucose (FSG) concentration and glycated hemoglobin-A1c (HbA1c) among the Chinese, Malay and Indian communities in Singapore. The study found a significant interaction between FSG and ethnicity on HbA1c. The correlation between FSG and HbA1c among Chinese subjects was found to be 0.25 (95% confidence interval [CI]:0.2–0.3) relative to the Malays (0.38, 95% CI: 0.30–0.45) after adjustment for age, gender, serum creatinine concentrations, body mass index, duration of diabetes, use of sulfonylureas, metformin, insulin, hemoglobin (Hb) and red cell indices ($p = .005$) (Dalan et al., 2013). There was no statistically significant difference between Indians and Malays with respect to the correlation between FSG and HbA1c.

Key milestones in project. This project was initiated and the dataset received in August 2010. The initial data analysis and manuscript writing was completed fairly quickly, and the paper submitted to *Diabetes Care* journal in October 2010. Unfortunately, the manuscript was rejected, and then after re-analysing and rewriting the manuscript, it was submitted to *Endocrine Practice* in April 2013 where it was accepted and published. There was a long period in between the first and second submissions, and with hindsight, perhaps this could have been shortened by requesting for more regular updates from the collaborator.

Significant highlights of the collaboration. The initial dataset that was provided in an Excel file included data on each patient stored in a separate Excel sheet within the Excel file. The subsequent amended database from the PI contained all the information in one single Excel sheet. Otherwise, the database was well designed and collected with the variables well defined and coded, and with only the necessary variables for the analysis provided. In addition, because the analysis involved separate regression models for the different racial groups, the 'bysort' command was used to simplify the analysis as shown below:

```
bysort ethnicity: regress hbaic serum_fasting_glucose
age gender duration _ of _ diabetes bmi su metformin hb
mcv mchc mch creatinine
```

A user-written command called 'predxcon' was also used to calculate the predicted values and 95% CIs after adjusting for key covariate in the final multi-variate model, as shown below:

```
predxcon hbaic if ethnicity==1, xvar( serum_fast-
ing_glucose) from(0) to(35) inc(5) adjust(age gender
duration_of_diabetes bmi su metformin basal_insulin
prandial_insulin basal__and_prandial_insulin hb
mcv mchc mch creatinine)
```

Key Learning Points

1. Learn how to document the key steps in a collaborative project, as well as to keep track of the milestones and outcomes using a project file.

2. Equip yourself with tools such as timesheets and software programming tricks to better manage multiple projects.

3. Ensure the consistency and reproducibility of your results via the use of syntax commands, log files and good documentation steps.

4. Identify the right mentor to help you get started with managing projects and learn how to maintain a good relationship with your mentor.

5. Avoid the common pitfalls in project management.

References

Berk, R. A., Berg, J., Mortimer, R., Walton-Moss, B., & Yeo, T. P. (2005). Measuring the effectiveness of faculty mentoring relationships. *Acad Med, 80,* 66–71.

Dalan, R., Earnest, A., & Leow, M. K. (2013). Ethnic variation in the correlation between fasting glucose concentration and glycated hemoglobin (HbA1c). *Endocr Pract, 19,* 812–817.

Pocock, S. J. (1995). Life as an academic medical statistician and how to survive it. *Stat Med, 14,* 209–222.

Thabane, L., Thabane, M., & Goldsmith, C. H. (2007). Mentoring young statisticians: Facilitating the acquisition of important career skills. *Afr Stat J, 4,* 123–136.

9

Managing Collaborations

9.1 Introduction

In the course of work, an academic biostatistician may get to collaborate with people from diverse backgrounds, from professors of medicine to junior doctors in training, medical students and other medical professionals such as physiotherapists and nurses. He/she may also collaborate with other statisticians as well. Although it is usual for a biostatistician to be involved in several concurrent projects, he/she may not be actively working on each one of them at the same time. In the words of a colleague, Assistant Professor John Allen (personal communication), this is akin to 'managing babies in a busy hospital nursery ... while there may be many babies present, only 3–4 may be awake at any time needing attention!'

It is important to understand the background of each collaborator so that work expectations can be managed very early on during collaborations. Institutional structures can also be placed to facilitate the collaboration process and make the experience efficient and pleasant for everyone involved. The topics covered in this chapter should count as one of the most important for the collaborating biostatistician. Research collaboration spans a wide range of activities, and this chapter starts with a section on how to leverage the most out of such collaborations. Suggestions are provided on how to provide biostatistical collaborative input in a large and complex health care cluster setting. A distinction is made between collaboration and consultation, and a separate process is described for the latter. The chapter then describes some of the common profiles of collaborators a biostatistician may encounter and provides useful tips on how to manage such collaborations effectively to produce a favourable research outcome in a collegial manner. Finally, suggestions are made on how a biostatistician can respond to some unfavourable requests from collaborators, such as data dredging. The chapter also deals with negotiating authorship and working with international collaborators.

9.2 Getting the Most Out of a Collaboration

It is important to understand that not all collaborations end up with a tangible academic outcome such as a publication or a successful grant application. As we can see from Figure 9.1, starting from an initial list of collaborations, most usually end up at least as presentations/posters at a conference with only a few eventually resulting in publications/grants. Some principal investigators (PIs) may not even come back after the initial analysis of the data (especially if the analysis did not result in what they wanted to see!). The actual numbers of resulting publications/grants can vary quite substantially depending on the nature and quality of the project as well as the research experience and motivation of the PI. From personal experience, out of 100 initial collaborations, 10 eventually end up as publications and 4–5 projects as successful grant applications. The point to note is that not all collaborations result in a successful research outcome, and it may not be entirely obvious from the start which are the productive projects. It is also better not to make judgements about a collaborator as the wise adage 'Do not judge the book by its cover' goes. Personally, this author has had the experience of initially not looking at a particular new collaborator favourably and thinking that the collaboration was not going to be fruitful, only to have a high-impact publication resulting from that work! Conversely, once a working relationship has been established between the PI and the biostatistician, future collaborations are usually

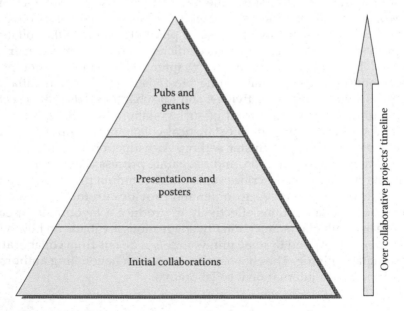

FIGURE 9.1
Pyramid showing typical collaborations that eventually lead to publications and grants.

productive, as both understand how to work with each other and the bio-statistician obtains a fair understanding of the clinician's medical area of speciality.

Sometimes, there can be signs that a project is not going to be productive. Examples include analysis of existing datasets with hypotheses crafted around existing data and variables without careful clinical justification, projects delegated to medical students and other medical trainees where the expected outcome is just a report or thesis (e.g. MBBS, MSc or MPH programme) or even a data analysis project where the PI has clearly stated interest in just presenting the results in a conference. There is a caveat here though: Some medical students can be extremely motivated, and with the right mentorship and close follow-up, there can be useful publications resulting from the collaboration. From the institution's perspective, collaborative work with the students may still be valued as critical work, and it is imperative then for the statistician to get recognition for the work done, in the event that publications do not arise from the work. It does not help that the student is usually not available for longer than the 1–2 years that he/she is usually allocated to undertake research, and the key is to keep in touch with the students even after they have graduated and constantly remind them of the project even as they get totally engulfed with clinical work in hospital postings. Another example of a possibly unproductive project is when the PI does not turn up for any meetings with the biostatistician, and instead delegates a research assistant or junior doctor to drive most of the research process. It is always a good practice to ask the collaborator at the beginning of the collaborative process about the intended outcome of the work as well as the anticipated date of submission for a publication or grant application (wherever applicable). This will provide a mechanism to remind the collaborator that a publication or grant is due.

It is a good idea to try and define the roles and responsibilities for each team member of the project early on so as to avoid duplication and make it easier for the PI to drive the process and ensure that the project comes to a fruitful completion. For instance, it is not uncommon for the PI to request the research assistant to perform some basic statistical analyses on the dataset. This may be because the person may be maintaining the database and has a fair understanding of the data, or it could be a measure to add more breadth to the research assistant's job scope to increase work satisfaction. Instead, by delegating this task entirely to the biostatistician, the research assistant's time can then be freed to focus on other activities such as systematic review of literature and the formatting of manuscript/tables/figures for publication. This may then free up the PI's time to do a bit more work to convert the poster or oral presentation (the intended initial outcome) to a complete manuscript, which can then be submitted for publication.

Sometimes, innocuous projects such as hospital routine data analysis as part of a hospital committee work or quality improvement projects can be converted into useful publications. For instance, the author was a committee member in Tan Tock Seng Hospital, Singapore, in 2005, tasked to examine the length of stay among inpatients by key diagnostic-related groups. One of the objectives involved comparing the length of stay between patients admitted on weekends versus weekdays, with an underlying belief that hospital service levels were reduced during weekends. The analysis was performed on routinely collected data from hospital administrative sources, as part of the Hospital Management Information System. Data on inpatient episodes was extracted with the assistance of the information technology department and organised by episode of care. After the report was completed and submitted to senior management of the hospital, it was decided that the project was conducted with enough scientific rigour and academically important to publish so that the methodology and results can be transferred to other hospitals. Subsequently, the research was published in the *BMC Health Services Research* journal, with the results showing that patients who were admitted on Friday, Saturday or Sunday stayed on average 0.3 days longer than those admitted on weekdays, after adjusting for potential confounders (Earnest et al., 2006). However, one should note that the use of administrative or routinely collected data for research has its limitations, including the unavailability of relevant and more detailed clinical data. For instance, in the abovementioned publication, we did highlight the limitation that there was the possibility of clinically more severe ill patients seen on weekends, but that we did not have enough clinical data to investigate this issue more thoroughly. Nonetheless, the key lessons in that collaboration include being able to alleviate any possible concerns about data (patient) confidentiality through the use of de-identified data, and to ensure that hospital administrators have an opportunity to look and discuss the results of the manuscript before it is submitted for publication.

9.3 Providing Collaboration in a Large Complex Institution – Hub-and-Spoke Model

Sometimes, it is possible that the biostatistician is part of a larger group of quantitative experts tasked to collaborate and provide support to a broad-based group of investigators from different medical specialities within a large tertiary institution or even across different institutions within a health care cluster or organisation. In such a scenario, it may be beneficial to set up institutional policies and other structures to enable effective and productive collaboration between clinicians and biostatisticians.

For example, the Centre for Quantitative Medicine (CQM), within the Duke-National University Singapore Graduate Medical School, Singapore, consisted of around 17 faculty members and 40 associate members, most of whom were biostatisticians and epidemiologists (https://www.duke-nus. edu.sg/research/centers/centre-quantitative-medicine-0. Date accessed: 15 April 2015). Among the several aims of CQM was to provide biostatistical and methodological support to clinician-scientists and other researchers from the nearby SingHealth cluster of hospitals and research institutes, including the Singapore General Hospital, the Singapore Eye Research Institute, the National Cancer Centre, the National Heart Centre and the National Dental Centre. CQM also allows for the provision of an academic home for quantitative experts dispersed throughout SingHealth system. It provides for collaboration with researchers in all SingHealth institutions using the hub-and-spoke model (see Figure 9.2). In such a structure, faculty members or associates are assigned to work with individual institutions. The allocation of time is based on need and use. However, CQM members are allowed to work with any investigator in any of the SingHealth institutions, although the aim of this model was to try to focus on relationships around specific interests and existing working collaborations.

The hub-and-spoke model has several advantages, including the following:

1. Flexibility in adapting to the biostatistical needs of individual institutions. Some institutions (particularly the larger ones) may require greater biostatistical input and collaboration than others. By housing all the statisticians under one roof, one may then

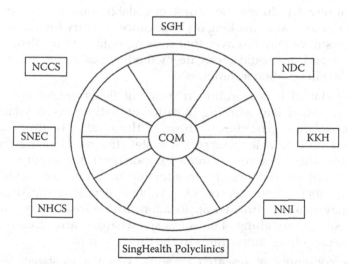

Note: CQM: Centre for Quantitative Medicine Singapore; SGH: Singapore General Hospital; NDC: National Dental Centre; KKH: Kandang Kerbau Hospital; NNI: National Neuroscience Institute; NHCS: National Heart Centre Singapore; SNEC: Singapore National Eye Centre and NCCS: National Cancer Centre Singapore.

FIGURE 9.2
Providing multi-institution collaboration using the hub-and-spoke model.

'farm' out biostatistician time (or full-time equivalent [FTE]) to each institution, depending on the needs. Some smaller institutions may also not have enough resources to recruit a biostatistician, and may then opt to pay for 0.5 FTE or equivalent, under the hub-and-spoke model. The arrangement can be formalised through secondments or project-based contracts depending on the nature and volume of projects. For example, such secondments were used to meet the needs of several institutions, including the Singapore National Eye Centre, the National Neuroscience Institute and the SingHealth polyclinics. It is a good idea to review these arrangements regularly (say annually) to see whether the needs have changed (e.g. institutions may well realise that they need 0.4 FTE instead of 0.2 FTE based on longer documented time spent on projects recorded in timesheets).

2. Allowing biostatisticians to specialise in a clinical speciality. By assigning biostatisticians to specific institutions, for example the cancer centre, the biostatistician is then able to specialise in statistical methods that are peculiar to that particular clinical field (e.g. biomarker diagnostic discovery for cancer, and psychometrics and quality-of-life scales for rheumatology). The biostatistician would also be more efficient in producing output because he/she will get quite familiar with the statistical models used in the field. Sometimes, it is possible for the biostatistician to develop independent methodological research during the course of collaboration in a specific field. For instance, after working on large cancer registry for a while, the biostatistician may discover that missing data are prevalent in the dataset and this could motivate methodological work around missing data imputation techniques.

3. Providing statisticians working in different fields opportunities to interact. Statisticians working in different institutions may face academic isolation, particularly those in the smaller institutions which may only be able to recruit few statisticians. Housing the statisticians under one roof provides opportunity for junior statisticians to obtain guidance and mentorship from the more senior faculty members in the centre. Ostensibly specialised statistical models may also have broader applications across institutions, for example, spatial modelling of out-of-hospital cardiac arrests can be applied to modelling cancer rates in the cancer centre.

4. Allowing continuity of research projects. When a biostatistician leaves employment or decides to undertake further studies, the gap in service can be temporarily bridged by other staff until replacement staff are hired and trained. This is acutely important because biostatisticians are usually short in supply, and it takes a relatively longer time to competently train a biostatistician.

9.4 Providing Consultations

It is important that a distinction is made between collaborations and consultations. The former involves repeated (and sometimes lengthy) contacts between the biostatistician and his/her collaborator, which often results in a publication or a grant submission with the biostatistician being named as a co-investigator/collaborator. The order of authorship and nature of involvement in the grant (e.g. co-investigator or collaborator, etc.) will need to be negotiated beforehand, but for substantial input in the project, the biostatistician is usually offered senior authorship (usually second or third) or named as a co-investigator in the grant submission. A consultation is usually brief (half-an hour to an hour) where the biostatistician provides advice on the statistical methods used, sample size calculation and even the design and methodology of the study. Usually, for the one-off advice, the biostatistician is not formally involved in the project, although occasionally an acknowledgement is provided in the manuscript. A consultation can be a more appropriate mode of support for a clinician who is experienced in research and who is able to handle the more basic statistical analysis required for the project using Statistical Program for Social Sciences (SPSS), but who would like an experienced biostatistician to review the analysis output. However, it is useful to draw up a process or SOP to provide such consultancy services, as the biostatistician may find himself/herself overwhelmed by such requests.

At CQM, consultancy clinics used to be run 1 day on a fortnightly basis, with each consultation session scheduled for an hour. The clinics were usually conducted by a pair of biostatistician and epidemiologist. The scope of service during the clinics included providing advice and guidance on study design, statistical analysis, research grant applications and scientific publications. The appointments were scheduled through an administrative assistant, and consultations were held in an office with a desktop, including all necessary statistical and related software. To ensure that the clinics were run efficiently and that they met the collaborator's expectations, the following work process was developed. Prior to the clinic session, the collaborator was requested to send a one-page summary of all the relevant information regarding the project to be discussed during consultation (including the context and expected outcome) and watch relevant Voice-Annotated Powerpoints (VAPs) relating to the request for consultation. Figure 9.3 shows the expected workflow, including the specific VAPs to watch and the critical information to provide the biostatisticians before attending the consultancy clinic. Common reasons for consultation were classified into 'sample size calculation', 'data analysis' and 'statistical analysis plan for grant application'. For instance, if a collaborator would like to calculate the sample size for a grant application, he/she would be requested to watch VAPs on the various components that go into a sample size calculation (e.g. type 1 and 2 errors, effect size and pilot data) as well as hypothesis testing. It is anticipated

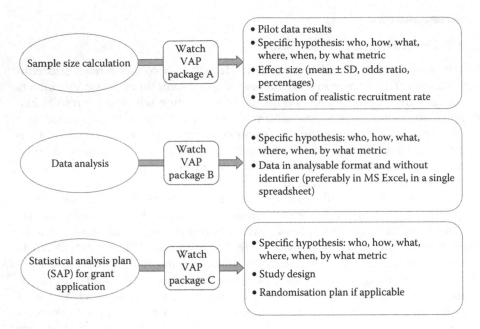

FIGURE 9.3
Workflow of VAPs to watch and critical information to provide prior to consultancy clinic.

that after watching the videos, the terms and phrases used in sample size calculations would be better understood, and the collaborator would then be able to bring along all relevant data and information to the consultation clinic so that the consultation session can be more productive. Expectations can also be better met by hopefully providing the sample size numbers on that day itself instead of requesting for more information during the meeting and unnecessarily delaying the process.

The other area in particular where collaborators may benefit from information in advance is in database preparation. Materials on how to ensure the database they have created is 'ready' for analysis should be sent to collaborators. From experience, it is always amusing to see how collaborators think that they have a ready dataset, when in reality it is far from ready. For instance, it is not uncommon to receive an Excel database from a collaborator, which has alphanumerical values in the fields and subgroups differentiated by different colour codes! In Chapter 3, we discussed some of the common mistakes made in database development, and also talked about how to ensure that the database is properly and efficiently set up. A final note about consultations: If the initial consultation has been a positive experience for the collaborator, and if further substantial input is required for the project, the consultation may even evolve into a full-fledged collaborative project.

9.5 Seven Faces of Collaborators

An eminent medical statistician has mentioned that 'doctors vary enormously in their personalities, appreciation of statisticians, ability to do research and the importance of that research' (Pocock, 1995). In a similar vein, a broader classification of types of collaborators that a biostatistician may encounter during work is described here. The aim is to equip biostatisticians and help them better manage the expectations of their collaborators by understanding their profile. The classification is by no means exhaustive and is based on personal observations of the author and colleagues he has spoken with, and it should be mentioned that collaborators can evolve and move across classifications after working relationships have been established, and they become more experienced, and hence the classifications are not static.

9.5.1 The Auto-Pilot

The auto-pilot collaborator is the (almost!) perfect collaborator who understands how the research process works and is experienced in working with biostatisticians and other research personnel. He/she comes to the initial meeting with very clearly defined aims and hypotheses. If looking for a biostatistician to collaborate in an analysis for publication, he/she will come with a cleaned and coded spreadsheet that is 'ready' for analysis. There would generally be one meeting to discuss the scientific elements of the project (e.g. aims/hypothesis, statistical analysis plans) along with timelines (e.g. how long the initial analysis will take and anticipated date of submission of manuscript) and the scope of work for each member of the collaboration (e.g. biostatistician to perform power calculations, statistical analysis and contribute towards the writing of the manuscript, PI to write up the draft of the manuscript, etc.). There is usually a minimum number of re-analysis as most aims are determined *a priori*, and the collaborator is quite clear in what is needed. There would be no issues about including the biostatistician as a co-author in the manuscript, although the order of authorship will be finalised based on the amount of contribution to the project *vis-à-vis* other collaborators. The manuscript is usually submitted in a prompt manner to a journal for publication and, if rejected, revised for re-submission to other journals until it has finally been accepted. Along the way, if the project has been presented in conference proceedings, the biostatistician would be informed and acknowledged. In essence, during such collaborations, there is minimal input from the biostatistician with maximal output gained. This productive collaboration would usually strengthen the working relationship and result in further collaborative projects between the two parties. Generally, more experienced researchers who have published extensively fall under this category of collaborators.

Clinician-scientists and other researchers who have been given protected time for research may also fall into this category. It may be possible to identify such collaborators by talking to other biostatisticians who have worked with them earlier. However, it is difficult to prospectively identify such collaborators until one has actually worked with them.

9.5.2 The Pseudo-Statistician

The pseudo-statistician is a clinician who has always wanted to work as a statistician as a second profession. Typically, he/she may have attended some statistical software classes, for example SPSS in the past, and feels that he/she is sufficiently equipped to undertake all the necessary statistics for the project. However, knowledge of using SPSS may not equip a person in terms of understanding the appropriate statistical test to use for the data. For instance, although one may know how to perform a Student's t-test, running the t-test to look at the association between gender and mortality (coded 1 for dead and 0 for alive) is not going to provide the correct p-value! A typical request for assistance from a pseudo-statistician goes something like this: 'I have performed all the necessary statistical tests for the manuscript. I just need some advice on whether I have done is correctly. 5–10 minutes of your time is probably enough. Can you help?' The reality is that the consultation is probably going to take much more than 10 min. As any experienced surgeon will tell you: it is easier to perform the actual operation, rather than having to fix someone else's botched job. The pseudo-statistician also has a habit of equating large amount of statistical analysis output with research productivity and may not understand multiplicity issues such as data dredging or multiple comparisons. This collaborator also has a tendency of wanting to employ complex statistical techniques in the analysis just because someone else has used them before and the tool sounds fanciful!

9.5.3 The 'Harry Houdini'

This describes collaborators whom you get to meet a couple of times before they disappear from your life altogether. These collaborators are not really interested in a publication but want to submit an abstract for a poster or oral presentation for a conference proceeding. Typically, the biostatistician is approached close to the deadline for submission of abstract, often with a false impression that not much statistical work is required for the abstract submission, and sometimes the initial statistical analysis may have been performed by the collaborator. The collaborator may also have a mistaken notion that the final poster/presentation can have very different results from that originally submitted in the abstract, so the work can be done in a hasty fashion and does not need to be 'exactly correct'. Consequently, the collaborator is not seen after the poster has been presented. To pre-empt such situations, it is useful to request for the outcome of the collaboration during initial meetings. If it is just for a conference presentation, the project may be turned down or channelled to other service provisions such as a consultancy clinic run by the biostatistician. For collaborators who try and circumvent the problem by claiming that the work will 'eventually be published', it is useful to ask them for an anticipated date of submission of manuscript during the initial meeting. Closer to the date, the collaborator can then be 'chased' for updates on the publication.

The other category of collaborators who tend to do the 'disappearing act' are those who do not have any interest in conducting research in the first place. This includes junior clinicians or trainees who have been asked by their supervisors to look for a biostatistical collaborator, sometimes after having presented at a departmental meeting. Grant proposals that have been knocked down because of a statistical critique may also fall within this category. Sometimes, the manuscripts or grant proposals are largely completed by the collaborators without any biostatistician's input, and the collaborators may not feel that substantial input from a biostatistician is required. In these instances, it is likely that the biostatistician may not be included as a co-author when input is provided. Instead, an acknowledgement is usually provided in return for the input. It may be possible for a biostatistician who has provided input for a publication to be acknowledged in the publication without having seen the final publication. This generates problems in terms of accountability. For instance, incorrect statistical analysis may be attributed to the biostatistician. It is advisable for the biostatistician to negotiate for authorship early on during the collaboration, and Section 9.10 provides advice on how to achieve this.

9.5.4 The *p*-Value Hunter

The *p*-value hunter is usually full of ideas and hypothesis and is overly optimistic about the project. The collaboration may start with the usual three to four aims and hypotheses, but when results are not significant, the collaborator may then request for subgroup analysis or formulate additional hypotheses to be tested. Sometimes, a request is made for the analysis to be restricted to particular subgroups of patients or even collecting more samples (with sequential analysis to be performed without any adjustment) with a goal to achieve statistical significance. The inexperienced biostatistician may feel intimidated to follow these instructions and perform the additional tasks, even though he/she is not agreeable. Attempts to inform the collaborator about the potential perils such as multiplicity may be futile. To circumvent such a problem, a mutually agreed statistical analysis plan should be drawn up at an early stage during the collaboration, and concerted efforts should be made to stick to the agreed plan. The collaborator should also be informed about the process of developing an impactful research question and how new studies can be adequately and properly planned by making sure the aims and hypotheses are clearly drawn up, and to ensure that the study is sufficiently powered to detect any clinically meaningful effect size in the key outcome measures. Section 9.8 provides strategies on dealing with related issues, including data dredging.

9.5.5 The Sceptic

The sceptical collaborator is one who is not easily convinced with the statistical advice or input provided by the biostatistician. The issues may not necessarily relate to the choice of statistical techniques to apply for specific variables as these are pretty much straightforward once the aims and hypotheses are described in details along with the nature of the variables collected (e.g. *t*-tests to compare means in two groups, and Chi-squared test to compare proportion in two groups). Instead, topics are usually related to strategies or approaches used in multi-variate modelling (e.g. variable selection methods), the appropriate use of transformation on skewed data or decisions around handling of missing data in a project. For example, the biostatistician may have suggested a logarithmic transformation on the outcome variable, length of stay, due to skewness in the distribution observed when plotting the histogram. However, the collaborator may have read of a different strategy in a related journal article (e.g. categorising the variable). To clear this confusion, sometimes, the collaborator may approach another biostatistician or even look up resources available on the Internet. Usually, when the second opinion concurs with the initial advice, the collaborator returns back to the original biostatistician. If they differ though and the latter advice suits, he/she may take it on and perform a 'disappearing act' with the first biostatistician. The other potential area where a collaborator is likely to seek a second opinion is around sample size calculation. When the initial sample size calculation results in an excessively large number, which may not be practical or costly to collect, the collaborator may visit another biostatistician with a 'revised' set of parameters for sample size calculation with an expectation of obtaining a smaller number. Although the number of collaborators seeking the second opinion for biostatistical input is expected to be small, personal experience has highlighted at least one instance where a collaborator had sought my judgement for a statistical advice provided by my colleague. Although I suggested a different statistical approach from the initial advice, this was done in a meeting with all three of us present, so that a robust discussion can be held, and there is a mutual understanding about how the optimal approach was selected.

9.5.6 The Passive Collaborator

The passive collaborator basically describes a person who is not really interested in research. He/she may not have had any training (formal or informal) in designing and conducting research (e.g. through courses on research methodology, database design, statistical methodology, etc.), and could be primarily heavily involved in the delivery of patient care in the hospital. His/her foray into research could have started with a gentle prodding from supervisors, sometimes with a condition of promotion linked to publications. Such publications can be termed key performance indicator publications. These can be based on poor-quality datasets; sometimes collected in the clinical departments for purposes other than research (e.g. clinical audits or quality improvement projects). Typically, the collaborator attends the meeting with the biostatistician with the data in an Excel file, because Excel is the most commonly available database tool in the hospital. The database is usually not adequately organised or coded, and can be riddled with incorrect or inconsistently entered data. The collaborator has a single aim in mind: that is to secure a publication. There are no pre-determined research questions, aims and hypotheses. The collaborator may not have read the literature adequately to define the important questions and is happy to 'discuss' with the biostatistician on 'what can be realistically analysed' in the available dataset. The collaborator is happy to take any suggestions from the biostatistician on what can be possibly analysed from the dataset to derive a publication, a process termed data-driven analysis rather than a hypothesis-driven process. Such projects are usually not productive and do not yield favourable outcomes. If one is lucky, a publication may eventually actualise, albeit after a long period of time.

For projects such as these, one suggestion is to explain to the collaborator the process of research leading to a publication, which should ideally start with a careful discussion of the research question, followed by the aims and hypotheses, study design and statistical methods. The time commitment for undertaking a research project should also be emphasised along with a clear demarcation of roles and responsibilities of each personnel in the collaboration. It would also be useful to suggest that a more senior clinical colleague (who may be more experienced in research) be involved in the project to mentor the collaborator and to explain potential limitations of the current dataset, and maybe even explore the possibility of obtaining more relevant or comprehensive data from other sources to answer the research questions.

9.5.7 The Faceless Collaborator

The faceless collaborator is someone who is extremely busy (usually a clinician who is heavily committed to clinical duties) to meet up with the biostatistician. One would get a request for collaboration through the e-mail or telephone and a commitment to meet up to discuss the project, only to have the meeting postponed because of last-minute reasons such as an unexpected call for surgery or delay in seeing patients in the ward.

This is unavoidable but may be repeated once too often. Sometimes, a representative or proxy (e.g. a junior registrar or even a medical student) is sent along with the dataset to be analysed and an apology from the PI. The representative may not be well versed enough with the project (especially the aims and hypothesis) to have a productive discussion and draw up a meaningful action plan with the biostatistician. This author has previously met with research assistants who were sent to discuss the statistical analysis plan by the collaborator who was apparently busy in clinic. Such meetings are futile as the research assistants are usually not able to provide answers to key questions, which would help determine the analytical plan for the project. The usual response would be 'I'm not sure what is the main hypothesis and outcome measure here. I will need to check with professor XXX and get back to you'. The collaborator was also absent during subsequent scheduled meetings, and as expected, the project did not have any tangible or meaningful outcome.

One lesson learnt was to make sure that the PI attend (at least the initial meeting), rather than sending someone else on his/her behalf. It is also a good idea to request for a summary of the study (including aims and hypotheses) before meeting up. Clinicians sometimes are busy with clinical duties and have ward rounds, so the biostatistician should be flexible about arranging meetings outside of normal working hours, and perhaps work out a staggered working arrangement with his/her supervisor.

9.6 Useful Strategies to Adopt in Successfully Managing a Collaboration

1. Where possible, try and provide a context to the problem early in the discussions with your collaborator (Chatfield, 2002). Specifying the scope of work, defining the research question and formulating the aims and hypotheses are important preludes to the research methodology and statistical analysis plan, and these should be firmed up sooner rather than later to avoid frustrations on both sides. This would usually entail listening carefully to the clinical description of the problem and understanding the processes involved.

2. When asked to perform an analysis that you know is incorrect, do not be afraid to say no. However, provide reasons for the decision. Where possible and without a compromise on the scientific quality, offer an alternative solution. For instance, if the collaborator is not keen on logarithmic transformation on skewed data due to interpretation issues, consider suggesting centring the variable around its mean value or even non-parametric models.

3. Always make it a point to read up on your collaborator's research topic before meeting up. It helps with the discussion process and also lets your collaborator know you are interested in the field. If possible, have a copy of a related journal article in front of you during the meeting.

4. Try not to get involved in projects at the last minute. Politely decline and offer reasons (e.g. you are busy with other projects). Advise collaborators to approach you earlier for future projects, so that this does not keep on happening. Building up a strong relationship with the collaborator through an early introduction and constant interactions will help build a good relationship and win over the trust of the collaborator who may then be more willing to listen to your advice and treat you as a person of authority.

5. Do not hesitate to ask the collaborator to slow down or repeat information that you do not understand. It is better to do so early during the conversation. It may be useful to make reference to figures and illustrations to highlight concepts. Abbreviate medical terms that you are unable to pronounce clearly.

6. Foster mutual respect. As described by Don Berry in the *Amstat News* magazine (http://magazine.amstat.org/blog/2012/02/01/collaborationpolic/. Date accessed: 17 June 2015), 'Respect yourselves, including by establishing a statistical reputation outside of your institution. Be confident, but keep learning, especially about science and medicine. Seek to achieve scientific and medically important goals. Think outside of your box. Listen more than you talk'.

7. Be proactive in choosing your collaborator rather than waiting and letting someone find you. If possible, choose a collaborator who is working in a disease field that you may have a personal interest in (e.g. oncology or cardiology). Choose collaborators who have demonstrated that they are able to communicate effectively with a biostatistician and have no troubles with treating a biostatistician as a partner in research rather than a number cruncher or a p-value generator!

8. Develop a positive attitude in thoughts and use of words with the collaborator. Avoid having thoughts such as 'I'm the statistical expert and he/she should listen to me' or 'I should not bother explaining the statistical analysis plan to him as he/she will not understand', as these discourage collaborative discussions.

9. 'You don't have to like the person you are working with in order to create a workable working relationship' (Organisational Issues in Biostatistical Consulting/Collaboration. Professor Jay Herson, Johns Hopkins Bloomberg School of Public Health, Baltimore, Maryland).

10. Try to end unproductive collaborations early. It has been reported that 85% of biomedical research is wasted because of inadequate production and reporting of research (Enhancing the QUAlity and Transparency Of health Research. How much money do we waste on research? http://www.equator-network.org/2014/08/12/how-much-money-do-we-waste-on-research/. Date accessed: 21 July 2015). Some individuals or projects may not be feasible or productive, and it is better to end the collaboration as early as possible rather than risking the chance of squandering valuable resources on the project.

9.7 Responding to Unreasonable Work Requests

Many research projects initiated by clinicians in the hospital are not funded. This means that the clinician has to do most of the work himself/herself or with limited support, including applying for institutional ethics application, data collection and all other aspects of the study. Some clinicians may even expect statisticians to perform some level of data entry and data formatting work prior to the data analysis. These tasks involve a substantial amount of time (e.g. at least half a day's work); then clearly, this is not an efficient use of the statistician's time and should be avoided. The clinician should be advised to get a part-time research assistant to perform tasks such as data entry. However, the statistician should provide a supervisory role to the research assistant (specifically in terms of database design and coding of variables [covered in more details in Chapter 3]). Some tasks such as recoding new variables from existing ones in the database (e.g. recoding age into age groups), generating new variables (e.g. logarithmic transformations) and imputing missing data are best undertaken by the statistician, but the division of such work scope needs to be clearly communicated with everyone.

The statistician should also take the opportunity to work with the research assistant to make sure that the dataset only has the minimum number of variables required for the study and that they are cleaned and coded appropriately. This will make subsequent work easier for everyone! Sometimes, in the process of cross-checking the dataset, the statistician may discover errors in data entry (e.g. cross-tabulating gender with pregnancy and obtaining entries for males!). An updated and corrected database should be requested from the clinician. Alternative, the corrected data can be imputed in the dataset with comments documenting the changes included in the corresponding software programme file (e.g. Stata 'do' file) and email correspondences with the clinician saved for future reference. The roles and responsibilities of the statistician *vis-à-vis* other members of the study (e.g. research assistant, clinical research coordinator, etc.) should be clarified, preferably during the initial meeting with the clinician. Collaborative projects need clear delineation of roles and responsibilities, and this can be challenging when a clinician/collaborator is relatively new to the publication process and does not really understand the process. These issues need to be clarified and clearly defined and mapped out for all the collaborators during the first meeting. Unless the statistician is the PI and first author in the study, other requests such as systematic literature review, creating/formatting references for manuscripts and creating posters and presentation slides should be politely declined. It should be noted that the process of redefining roles and responsibilities is ongoing as this may change over time (e.g. the first author may not be able to work on the manuscript anymore, and it may then be necessary to get some else in the team to drive the process of completing and submitting the manuscript).

9.8 Reasoning with a Collaborator Who Engages in Data Dredging

It is not unusual for hospital-based researchers to have access to vast amounts of medical records for research, and this usually results in a database with a large number of variables, some of which may not be relevant for their research. It is then tempting for some collaborators to request for an 'all possible' statistical analysis plan, in order to trawl through the data, and hoping to discover some significant findings. The dangers associated with such data dredging and fishing expeditions for p-values that are significant and the associated problems with multiplicity in hypothesis testing have been discussed in Chapter 5. This should be communicated with the clinician. To pre-empt such requests, the statistician should discuss with the collaborator and agree on a fixed number of aims and hypotheses for the project, along with an associated statistical analysis plan before the analysis of the data. The statistician should not easily succumb to requests for additional variables to be included in the analysis or for unplanned subgroup analyses to be performed. Questions such as 'What is the clinical rationale for including this variable?' and 'Where is the evidence in prior literature that this variable is important in our analysis?' should be asked to discourage data dredging. Possible problems with the additional variables, such as excessive missing data and measurement errors and biases, can be highlighted to discourage their inclusion in the analysis. However, some minor changes or amendments to the original analysis plan can be accommodated after the initial analysis (e.g. testing for interaction and discovering the need to present data by subgroups, combining categories across well-defined groups due to small numbers, etc.).

9.9 Coping with an Unreasonable Request on Turnaround Time

Most requests for collaborations requiring statistical analysis leading to publications are relatively not urgent and come with a reasonable lead time, and hence allow for a realistic time for statistical analysis compared to grant proposals. Sometimes, there is a genuine reason to get a paper published really quickly (e.g. a competitor is working on a similar topic and is about to complete the project, there is an outbreak of a new infectious disease and it is important to quickly publish details of the illness such as clinical course of disease, treatment effects and disease outcomes). The statistician should prioritise projects according to their importance and urgency. In the event that the current project cannot be worked on due to lack of time, this should be communicated to the clinician as soon as possible, so that an alternative statistician can be sourced for the project in a timely manner. Depending on the number of projects that the statistician is currently working on as well as the complexity of the projects, the average turnaround time for an initial analysis of the data can vary from 1 to 10 days. In order to minimise miscommunication and better manage the expectations of the collaborator, the timeline for data analysis and other deliverables should be discussed, mutually agreed upon and clearly documented, preferably during the initial meeting. Some collaborators (especially those who are new to the publication process) may not understand the nature of the publication process and even underestimate the time it takes for a manuscript to get published (i.e. preparation of the manuscript for publication, peer-review and possible revisions to be made to the manuscript and rejections will cause further delays). Collaborators should be advised to approach biostatisticians well in advance, and not close to deadlines.

For the occasional collaborator who does not know the tasks involved in the data analysis process and demands for immediate and quick attention to his/her data, one possible solution includes informing him/her: 'I'm currently working on Professor X's (or someone important) project, and if you like, you may approach Prof. X to convince him that I would need to postpone working on his/her project, and that your project is more urgent'. Additionally, one may also turn down requests for collaboration on the basis that they are practically not feasible (e.g. too short a time to analyse the data accurately) or scientifically compromised (e.g. no adequate time allocated in the project for hypothesis generation and data analysis).

Compared to publications, requests for collaboration for grant applications have more immediate deadlines. This is mainly because grant calls are often fixed in timing across the year, and there is a short window between the grant call and the closing date. For instance, the project grant call for 2015 under the Australian National Health and Medical Research Council was opened in the online Research Grants Management System on 14 January and closed on 18 March. In addition, most institutions and hospitals have their own internal deadlines for review by their internal research office/review process, which may be 1–2 weeks (if not more) earlier than that date. Therefore, if a clinician waits until a grant call is open before thinking about a research proposal, there is often very little time to work on the proposal, get collaborators involved and develop a good proposal with well-written aims/hypotheses, study design and statistical analysis plan. To compound the problem, demand for statistical collaboration in a grant proposal is also seasonal, with the statistician receiving several such requests at the same time. To mitigate the problem, reminders can be sent around to remind collaborators to engage statisticians early on, possibly even before the official launch of the grant calls. Rules and policies can also be set in place to ensure that there is a minimum time before grant submission closing date for collaborators to approach the statisticians. Proposals can also be pre-reviewed and sieved to ensure that they are of a sufficient quality and have well-developed aims and hypotheses to benefit from a biostatistical collaboration. Sometimes, the collaborator requests for a quick sample size calculation at the last minute, and it is actually fine to 'refuse', as one researcher did, quoting that 'Sample size is too integral a part of the design itself to be patched in at the end of the process' (Wittes, 2002).

9.10 Negotiating Authorship

It is important that the biostatistician discusses with the collaborator about authorship issues early on during the project. This is to ensure that authorship expectations are managed and misunderstandings do not occur later on during the project. For his/her input in the project, the biostatistician should be included as a co-author in the manuscript. This is a form of accountability so that when journal reviewers query on the statistical methods, the biostatistician has the mandate to respond to the questions raised.

Sometimes, it may be difficult to negotiate authorship with the collaborator. Common reasons given include the following: too many authors are already listed in the paper, other authors have objected to including a biostatistician as a co-author or it is sufficient to acknowledge the biostatistician in the manuscript, in lieu of authorship. The International Committee of Medical Journal Editors (ICMJE) has developed a series of guidelines and recommendations to review best practices and ethical standards in the conduct and reporting of research in medical journals. It has a set of recommended documents for authors and journals listed on their website (http://www.icmje.org/. Date accessed: 29 April 2015). It includes the Recommendations for the Conduct,

Reporting, Editing and Publication of Scholarly Work in Medical Journals. In one such document that describes the role of authors and contributors, the ICMJE lists the following four criteria, all of which need to be met to be considered for authorship:

1. Substantial contributions to the conception or design of the work; or the acquisition, analysis or interpretation of data for the work
2. Drafting the work or revising it critically for important intellectual content
3. Final approval of the version to be published
4. Agreement to be accountable for all aspects of the work in ensuring that questions related to the accuracy or integrity of any part of the work are appropriately investigated and resolved

Clearly, a biostatistician who has contributed to the analysis of the data and has provided input in the drafting and finalising of the manuscript will be qualified for authorship based on the abovementioned criteria. This guideline can be highlighted to the collaborator to successfully negotiate authorship. To avoid any sort of abuse of the use of the criteria to exclude biostatisticians from authorship (e.g. by not inviting the biostatistician to contribute to the writing of the manuscript), the ICMJE has clearly stated that those who meet criterion 1 should be invited to participate in the review, drafting and final approval of the manuscript. All individuals who qualify for the authorship must be listed on the paper. Those who do not qualify for authorship are listed in the acknowledgement section of the manuscript (e.g. data entry assistants, interviewers, medical writers, etc.). The guidelines are meant to be used for ICMJE member journals, and these are usually stated in the instruction to authors, and the authors are requested to explicitly state that they follow such guidelines. The order of authorship may vary according to the relative contributions of individual authors, and this is usually proposed by the first author and agreed upon by the other co-authors. The first, second, third and last authors occupy the important positions, and these usually include the corresponding author or the author in charge of submitting the article to the journal, and ensuring that all the relevant paperwork is completed and submitted. The PI is usually the first author in a collaborative project, unless the project is based on substantive statistical work for which the biostatistician takes the lead and is then the first author. It is also recommended that a written document describing the authors' roles and responsibilities and authorship order is prepared and circulated in advance (Kotz & Cals, 2013). This will help to circumvent any authorship issues that may crop up. The contents of the authorship may change over time (e.g. the co-author has decided not to be included due to insufficient time at hand to work on manuscript, and hence needs to be removed), but essentially the document allows for the assignment of responsibility to the co-authors.

Sometimes, the PIs may decide to attribute the contribution of the statistician in the 'acknowledgement' section of the manuscript. The final analysis and manuscript may/may not be seen by the statistician, and there is unfortunately no form of accountability in this process, and the acknowledgement may not even count as a tangible output for the statistician. It may also be possible that inappropriate statistics or graphs may be presented and entirely bypass the statistical review during a journal submission, just because a statistician was listed in the acknowledgement section! (Chatfield 2002).

9.11 International Collaborators

Occasionally, potential collaborators may not be within the same institution or country, and it may be necessary for the statistician to reach out further to collaborate with both clinicians and statisticians who may be working in the same field. Collaborating internationally helps to raise the profile and academic stature of the statistician. Geographic proximity and distance are no longer obstacles to collaborating with someone from another country or continent with the advent of technology, in particular video-conferencing facilities. Many universities have such facilities available on campus, and there is also free software available, such as Skype (http://www.skype.com/en/. Date accessed: 23 June 2015). Online file-sharing tools such as Dropbox are also available (https://www.dropbox.com/. Date accessed: 23 June 2015), where one can share the latest manuscript, proposal or other documents online with collaborators. Most of the time though, collaborations start locally and then continue interstate or across countries when one of the collaborators relocates to another institution in a different country.

Attending conferences is one way of getting to know others who are working in the same field, to present one's work and hopefully generate interest and collaboration as well as obtain feedback. If the work is statistical or methodological, then usually statistical conferences would be useful to attend. Otherwise, if the work is applied to a clinical problem, then the appropriate medical conference would be beneficial. Several key statistical societies that regularly host statistical conferences are listed in Sections 9.11.1 and 9.11.2.

9.11.1 General Statistical Conferences

1. American Statistical Association. http://www.amstat.org/meetings/. Date accessed: 17 June 2015.
2. Royal Statistical Society. http://www.rss.org.uk/RSS/Events/RSS/Events/Events.aspx/. Date accessed: 17 June 2015.
3. Statistical Society of Australia. http://www.statsoc.org.au/events/. Date accessed: 17 June 2015.

9.11.2 Biostatistics Conferences

1. International Society for Clinical Biostatistics. http://www.iscb.info/Events.html. Date accessed: 17 June 2015.
2. International Biometric Society. http://www.biometricsociety.org/meetings-events/events-calendar/. Date accessed: 17 June 2015.

Key Learning Points

1. Learn how to make the most out of a collaboration.
2. Create a framework to provide collaboration in a large and complex environment.
3. Differentiate between a collaboration and consultation.
4. Identify the seven faces of collaborators and see which one your collaborator fits in.
5. Acquire useful strategies in managing collaborations.
6. Negotiate to get recognised for your work (e.g. authorship).
7. Work with international collaborators.

References

Chatfield, C. (2002). Confessions of a pragmatic statistician. *Statistician, 51*(1), 1–20.

Earnest, A., Chen, M. I., & Seow, E. (2006). Exploring if day and time of admission is associated with average length of stay among inpatients from a tertiary hospital in Singapore: An analytic study based on routine admission data. *BMC Health Serv Res, 6*, 6. doi:10.1186/1472-6963-6-6.

Kotz, D., & Cals, J. W. L. (2013). Introducing a new series on effective writing and publishing of scientific papers. *J Clin Epidemiol, 66*, 66–67.

Pocock, S. J. (1995). Life as an academic medical statistician and how to survive it. *Stat Med, 14*(2), 209–222.

Wittes, J. (2002). Sample size calculations for randomized controlled trials. *Epidemiol Rev, 24*(1), 39–53.

10

How Not to Design, Analyse and Present Your Study

10.1 Introduction

There are many courses and books that one can consult to learn how to correctly design and analyse a study. It is equally important to know the 'how not to' as well. A simple analogy is to look at the way we take instructions on the labels on our shirts before putting them to wash (e.g. do not machine wash, do not tumble dry, etc.) if we want a shirt that looks nice and lasts longer! Grant review panels, for instance, often look for the single statistical or methodological flaw in the proposal, and this is enough to sink the proposal. The problem of sub-optimal design and analysis is prevalent in the various types of health care output, including health care reports, conferences and even peer-reviewed medical journals! Within each domain, the problems are also diverse. For instance, in a survey (Harris et al., 2009) among editors and statistical reviewers of 54 high-impact psychiatry journals to determine the statistical or design problems they encountered most often in submitted manuscripts, the following areas were identified: failure to map statistical models onto research questions, improper handling of missing data, not controlling for multiple comparisons, not understanding the difference between equivalence and difference trials and poor controls in quasi-experimental designs. Most of the mistakes in the design, analysis and reporting of studies are perpetuated from previous published studies using the same bad methods. However, as one eminent statistician put it, 'precedence is a justification for lawyers not scientists, and it is logic not precedence that has to determine the way we measure' (Senn & Julious, 2009) or design and analyse the study in our case.

This chapter is organised as follows: Firstly, issues related to incorrect use or application of study designs are covered, which include the choice of study designs, inadequate sample size calculation and improper use of randomisation. Next, the inappropriate application of statistical tests, including failure to check for the assumptions behind a statistical test, is discussed. Issues such as standardisation, mistaking association with correlation, over-fitting in regression modelling and data dredging are covered. Following the next natural stage of the research process, improper reporting and interpretation of data and results are highlighted, which include the presentation of tables and figures. Finally, the interpretation of study results, including the possibility of providing 'spins' in the writing up of results, is discussed. Examples are drawn from a wide range of sources, including journal articles, books, websites and conferences, as well as a related journal article by the author (Earnest et al., 2012).

10.2 Choosing the Inappropriate Study Design

The study design contributes a large extent towards the scientific validity and quality of a research project. It is difficult to correct and fix an incorrectly designed study unless one can travel back in time! The biostatistician should endeavor to get involved early during the collaborative project, ideally when the study is being planned, so that he/she may provide input in the design of the study. The choice of study design depends on a number of factors, including objectives of the study, prevalence of outcomes and risk factors and to a lesser extent available cost/logistics and timeline for project completion. In order to choose and apply the correct study design, one must understand the various designs and their associated features. Chapters 1 and 2 provide detailed discussions on the various features of the different epidemiological study designs available to a health researcher.

If the aim of the study was to establish the efficacy of a new drug compared to an existing drug, one may wish to conduct a randomised controlled trial (RCT), particularly if there are concerns relating to confounding and minimising selection/interviewer/assessor bias. However, if the objective was to identify a causal link between smoking and chronic obstructive pulmonary disease (COPD), a cohort study may be particularly useful. If the outcome was relatively rare (e.g. Buerger's disease), then a case–control design would be more applicable. Common mistakes include improper assessment of exposure (e.g. recall bias) in a case–control study, selecting cases and controls based on the exposure instead of outcome variable, using a cross-sectional study to establish causality and comparing groups from across different RCTs. Even within a single particular study design, incorrect procedures can be employed (e.g. selecting cases and controls from two different populations, selecting high risk cases, the healthy worker effect, etc.). Some of these examples have been discussed in more details in Section 1.5.

For example, in a case–control study to examine the relationship between diet and hypertension among Australians, investigators may have sampled hypertensive patients from a tertiary hospital in Melbourne and the controls from a community-based sample. This gives rise to the healthy worker effect bias, as the controls are more likely to be healthier, possibly with better dietary habits. The controls should have been selected from the same population as the cases. Selecting cases and controls based on the risk factor or covariate under investigation (e.g. dietary habits) is another common fault in designing a case–control study, with the identification of two separate groups giving one a false impression that a case–control design has been employed. Another related problem lies with 'survivor bias', where a segment of the recruited sample actually die before their outcomes are ascertained. For instance, it may be problematic to study the 6-month follow-up of the quality-of-life (QoL) scores among newly diagnosed advanced pancreatic cancer patients, particularly if the median survival is 3.5 months!

It is also possible that a study was initially designed as an efficacy study (i.e. comparing the effectiveness of drug A vs. drug B), but when the results are found not to be significant, one may be tempted to analysis the data as if it was an equivalence trial instead. As discussed in Chapter 2, the features and nature of hypothesis testing are different between the two study designs and should not be mixed up. In meta-analysis studies, it is also possible to inadvertently compare treatment arms against placebos in different trials, and this can be problematic as the patient risk profiles are different across trials. Even though they are individual RCTs, the nature of randomisation only balances patient demographics within a trial and not across trials. Some of the deficiencies in designing studies can be attributed to the fact that many of these studies are based on data that has been collected for purposes other than research (e.g. clinical management systems, clinical registries, quality improvement projects, etc.). The study design is then limited to the nature whereby data has been collected (e.g. often historical cohort, rather than concurrent or prospective studies, and the inability to conduct RCTs) and inherent limitations in the data (e.g. missing observations and possible bias). The implications of poorly designed studies are huge. Poorly designed studies are less likely to get funded or published, and even when they do get published, it has also been reported that when well-designed and poorly designed clinical trials of the same treatment are compared, the latter shows larger treatment effects (Altman, 1991).

10.3 Selecting Too Few Subjects in the Study

Recruiting larger than necessary subjects results in wastage in cost and resources, and may expose additional people to unnecessary risks inherent in the study. With really large sample sizes (typically in large population-based studies or hospital data), it is sometimes possible to obtain statistical significance for tests, but for results that are clinically not meaningful.

For example, Table 10.1 shows hypothetical data, comparing four clinical outcomes across two hospitals (each with a sample size of 100,000 inpatient episodes). As we can observe, with such a large sample size, even small differences in the outcomes are reported to be statistically significantly different between the two hospitals!

Conversely, when too few patients are recruited, such studies may not adequately demonstrate a statistically significant result. These are typically known as under-powered studies. Usually, power is set at 0.80. Lower power implies a higher type 2 error (i.e. failure to reject the null hypothesis when it is actually false). Table 10.2 highlights how sample size is related to the statistical significance of a study (or p-value). If too few subjects are recruited (i.e. $n = 50$ per hospital), the possibility of type 2 error is then high. We are unable to conclude that the nosocomial infection rate in hospital A is 10% (absolute) higher than in hospital B. With all else held constant, if the study size was increased 10-fold, we find that the p-value is smaller, and a further 10-fold increase in sample size to 5000 will result in a significant p-value of .002.

Under-powered studies are common in the medical field. For example, in a review of RCTs assessing clinical efficacy of treatments for adult rheumatic diseases published in English between 2001 and 2002, of the 86 negative or indeterminate trials, 49 (57%) did not report sample size calculations (Keen et al., 2005). Of these trials that did not report power calculations, *post hoc* sample size calculations revealed that only 10 were adequately powered. Sometimes sample size/power calculations are not provided in details to enable reproducibility, as observed from one conference presentation (see Figure 10.1). Therefore, it is not possible to verify the author's claims about the power in the study, and hence access the quality of the study design.

TABLE 10.1

Comparison of Key Clinical Outcomes by Hospitals

Clinical Outcomes	Hospital A	Hospital B	p-Value
Health care-associated *Staphylococcus aureus* bloodstream infections (rate per 100,000 bed-days)	1.01	1.03	<.001
Mean breast cancer surgery waiting times	12.5	12.7	<.001
Mean lung cancer surgery waiting times	10.5	10.6	<.001
Emergency department time (% leaving within 4 h)	85%	84%	<.001

TABLE 10.2

Comparison of Nosocomial Infection Rates by Hospitals

Hospital A (%)	Hospital B (%)	Sample Size	p-Value
60	50	50	.315
60	50	500	.178
60	50	5000	.002

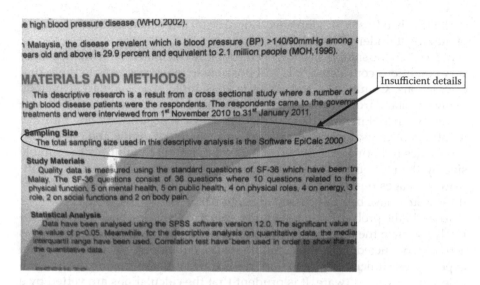

FIGURE 10.1
Insufficient details on sample size calculation.

Power calculations are sometimes done retrospectively after a study is completed, usually at the request of journal editors and reviewers, whenever the adequacy of the study size is in doubt. This is usually requested when the results from a statistical test are shown to be not significant (i.e. p-value > .05), but with a large enough effect size that is deemed to be clinically meaningful (e.g. hazard ratio of the magnitude 4). Studies that are underpowered are also less likely to get published as they do not provide a statistically significant result, which would then not warrant any attention from the journal editors or readers. This gives rise to the problem of publication bias in meta-analysis studies, which has an impact on the validity of the findings, regardless of the quality of the individual studies included in the review. Funnel plots are usually used to determine the existence of publication bias in studies (Egger et al., 1997).

Sometimes, sample size may have been performed, but because of inadequate allowances for dropouts and non-compliance as well as overestimating patient accrual, the actual numbers at the end of the study may be much lower, leading to under-powered analysis. The use of pilot studies or data from related research (e.g. a similar trial conducted by a colleague in the same department) may provide for realistic projections of patient accrual rates, dropouts and so on, which can then be incorporated in the planning of the sample size. Another related problem in under-powered studies lies in missing observations for key variables, including the outcome measures. Therefore, even if sample size calculation has been performed and all the patients have been recruited and manage to complete the study, if a substantial number do not provide response for the outcome or key explanatory variables, then this will again have a significant impact on the study in terms of making it under-powered.

Even when the sample size has been properly calculated with all the necessary conditions considered, it is still possible to make mistakes in the actual calculations. Common mistakes made in sample size calculation include using the standard deviation (SD) of baseline values, instead of the change values when looking for change in scores across time, undertaking multiple iterations of the sample size calculation (usually changing the effect size) until a desired number is obtained as well as failing to multiply the sample size by the number of comparison groups (particularly because most software packages report the sample size required per group or arm). Some of these issues have been discussed in Section 4.6.

To avoid the problem of having under-powered studies, one needs to accurately estimate the total number of subjects needed for the study before it is actually conducted. For simple sample size calculations, sometimes the more experienced clinician is able to undertake the calculations using freely available sample size software. It is prudent that the calculations are vetted by a biostatistician to avoid potential pitfalls discussed previously. The biostatistician can also help with sample size calculations in more complex study designs (e.g. cluster randomised trials, stepped wedge trials) and address other potential issues (e.g. dropouts, non-compliance, etc.).

10.4 Incorrect Use of Randomisation

Randomisation is an act of allocating patients to treatment or intervention arms in a manner in which the investigator has no control over. The aim of randomisation is to minimise selection bias and reduce confounding by ensuring that the groups are comparable in terms of key demographic and other important clinical variables. The key success of randomisation lies in the concealment of treatment (intervention) allocation from patients, investigators and assessors.

Asking the patient to 'randomly' choose between treatments A and B, or for an investigator to randomly assign patients to active and placebo groups based on how sick the patients appear, or to assign patients based on day/time of visit or odd/even room numbers are not valid forms of randomisation. Let's look at the hypothetical example of a geriatrician who wishes to evaluate the effectiveness of protein supplementation among elderly patients. His/her main outcome measure is hand grip strength. Suppose out of convenience, he/she randomises patients who present to his/her clinic on Mondays to protein capsules and those on Tuesdays to placebo. The study results can be potentially biased, given that senior citizens get a 2% discount for shopping on Tuesdays at the major supermarket chain! An equally serious problem with the randomisation process lies with the inadequate concealment of the treatment allocation from patients, doctors and those who are assessing the outcomes in the study. This may lead to potential bias in response and evaluating the outcome, as we discussed in Chapter 2. The randomisation list should be kept in a safe and secure location, and should not be printed and kept on the desk.

Some investigators also confuse between randomisation and random sampling. The latter is an act of selecting a smaller group of subjects (sample) from a larger population. There are various types of random sampling such as simple random sampling, stratified random sampling and systematic random sampling. Non-random sampling methods include convenience sampling and snowball sampling. Random probability sampling ensures that the results from the sample can be generalised to the population that it was sampled from. Some collaborators have a mistaken notion that a large sample size will most certainly provide a more accurate depiction of the population. Figure 10.2 shows the relationship between precision and accuracy. A study with a large sample size, but sampled in a non-random manner will provide us with a precise result, but unfortunately may not be accurate (Figure 10.2b). A smaller study with random probability sampling will provide relatively more accurate but imprecise results (Figure 10.2c). Ideally, one would want a large study with a randomly selected sample (Figure 10.2a).

Software such as Stata and Excel can generate random numbers or these can also be obtained from random number tables, often found at the back of elementary statistics textbooks. For instance, if one would like to select 10 numbers, each number randomly chosen between 1 and 100, one can type = randbetween (1,100) in a Microsoft Excel spreadsheet (Figure 10.3) to obtain the random numbers. Note that the numbers change when any action is performed within the Excel sheet, so it would be better to copy and paste the values into a new sheet and discard any formulae in the process.

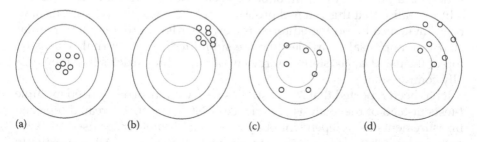

(a)   (b)   (c)   (d)

FIGURE 10.2
Accuracy and precision: (a) accurate and precise; (b) inaccurate and precise; (c) accurate and imprecise and (d) inaccurate and imprecise.

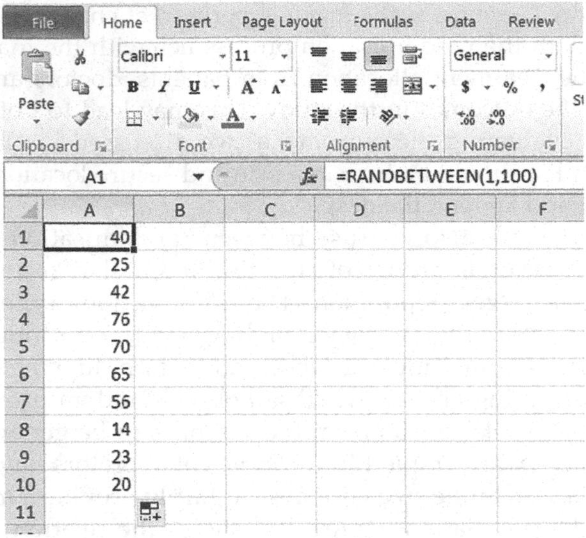

FIGURE 10.3
Generating random numbers in Microsoft Excel.

10.5 Undertaking Incorrect Statistical Tests

The type of statistical test one performs on the dataset depends on a number of factors, including the aim and hypothesis of the study, sample size, number of groups being compared and whether comparisons are performed within the same person. The nature of the data collected, the types of variables as well as the distribution of the variables are equally important. When in doubt, the clinician can consult a basic biostatistics textbook, although it is probably easier to approach an experienced biostatistician! The incorrect application and interpretation of statistical techniques in medical journal articles is a problem that cannot be neglected or trivialised. In fact, journal editors realise that this is a real problem, and increasingly statisticians have been invited to be on the editorial review boards to provide more regular statistical appraisal of articles that are submitted to the journal. This is on top of the usual *ad hoc* statistical review of individual articles that they usually commission.

It has been reported that 24% of studies reviewed incorrectly used multiple *t*-tests instead of the analysis of variance (ANOVA) test, 5% treated repeated measurements as independent observations and another 5% used the Chi-squared test when the expected frequencies was too small (Altman, 1991). In that same review of journal articles published in *Arthritis and Rheumatism*, it was found that 66% of the articles had at least one statistical error in them. Statistical errors in medical journal articles can be extrapolated to other fields, and even journals with a high impact factor are not immune to this problem. There are plenty of examples, not just in journal articles, but in medical conferences and other health-related reports. In one such health care conference, the poster in Figure 10.4 was observed to have presented mean values for two groups, but used the Chi-squared test incorrectly to calculate the *p*-value. I am fairly confident that the incorrect use of statistical analysis methods is a prevalent issue in other medical conferences as well.

Yet another mistake statisticians (and other researchers) make is to ignore dependencies in the observations and to instead treat them as independent observations. This is particularly in ophthalmology studies, where either one of the eyes or both eyes of the same patients are included in the study. Ocular measurements can be more alike within the same individual's eyes than between eyes from different individuals, and this could be due to structural similarities, genetic pre-disposition and so on. Traditional methods of analysis include choosing the best/worst eye, taking the average of both eyes or randomly selecting one eye. Other methods include analysing data from both eyes while treating them as independent samples, as well as analysing both eyes, but accounting for the intra-person correlation through a more complex model (e.g. random effects model). The last method is probably the most appropriate as it utilises most of the information that is available, but may not be popular, probably because some statisticians may not be familiar with methods such as multi-level models or generalised estimating equations that are typically used to analyse such complex data.

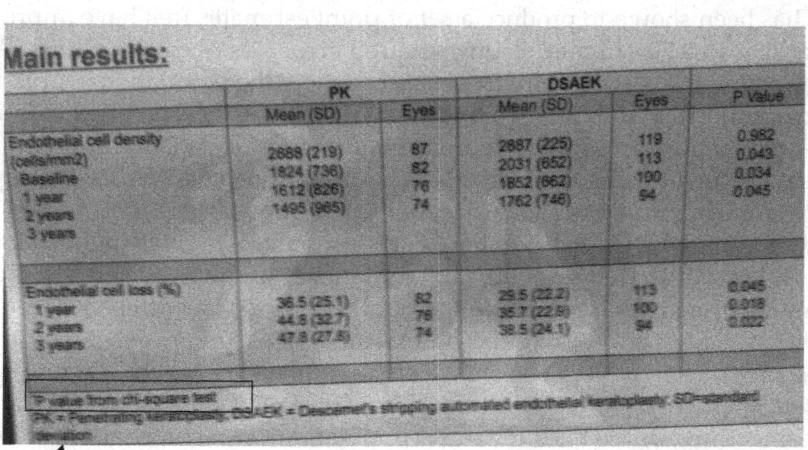

Main results:

	PK		DSAEK		P Value
	Mean (SD)	Eyes	Mean (SD)	Eyes	
Endothelial cell density (cells/mm2)					
Baseline	2688 (219)	87	2887 (225)	119	0.982
1 year	1824 (736)	82	2031 (652)	113	0.043
2 years	1612 (826)	76	1852 (662)	100	0.034
3 years	1495 (965)	74	1762 (746)	94	0.045
Endothelial cell loss (%)					
1 year	36.5 (25.1)	82	29.5 (22.2)	113	0.045
2 years	44.8 (32.7)	76	35.7 (22.9)	100	0.018
3 years	47.8 (27.8)	74	38.5 (24.1)	94	0.022

P value from chi-square test
PK = Penetrating keratoplasty, DSAEK = Descemet's stripping automated endothelial keratoplasty, SD=standard deviation

Incorrect statistical test

FIGURE 10.4
Incorrect use of the Chi-squared test.

When health data are organised in time intervals (e.g. weekly or monthly counts), there is the possibility of temporal correlation in the data. For instance, dengue fever (DF) notifications in Singapore have been shown to exhibit both seasonal trends (e.g. regular peaks around June or September) and epidemic trends (e.g. during the 2005 epidemic) (Earnest et al., 2012). Seasonal trends in DF can be attributed to climatic variables such as temperature and precipitation. Ignoring the autocorrelation in the data can result in artificially small standard errors in the model and give rise to spurious statistically significant results in the study. Autocorrelation in the data can be detected through correlogram plots, and when present, suitable time series models such as the autoregressive integrated moving average should be used to analyse the data.

A similar problem also exists when health data are presented in choropleth maps. This is often in the form of relative risk estimates of disease at an areal level (e.g. local government area or postal areas), which may or may not be standardised for demographics such as sex and age across the state or nations. It is known that ignoring geographical correlation inherent in the data may lead to a biased and inefficient inference, as the observations are strictly not independent (Earnest et al., 2007). This is particularly problematic in regression analysis when one intends to study factors associated with risk of disease at the areal level (e.g. associations with areal indices of socio-economic disadvantage). One possible approach is to use the conditional autoregressive (CAR) model to analyse the data by including both spatially structured and spatially unstructured random effects terms in the analysis to account for spatial correlation. This approach has been shown to offer a trade-off between bias and variance reduction of the estimates, and has been shown to produce a set of point estimates that have improved properties in terms of minimising squared error loss (Carlin & Louis, 2000). The CAR model is also useful in instances where the population in areas are sparse, and the model can then 'borrow strength' by incorporating information from nearby areas when there is geographical correlation present. The model can be fit using freely available software such as WinBUGS (http://www.mrc-bsu.cam.ac.uk/software/bugs/the-bugs-project-winbugs/. Date accessed: 2 July 2015). Health atlases are also increasingly being developed and displayed on websites by health authorities and other institutions as a means to provide information to the public and other researchers about the health of the region. Most of the time, simple rates are provided, and these are sometimes standardised for key demographics. However, the rates presented in such maps may not have been accounted for spatial correlation, and smoothing of the rates for small population counts may not have been performed. For example, the Health Quality & Safety Commission's Atlas of Healthcare Variation displays the rates of child admissions due to asthma or wheeze (and other health indicators) at the district health board geographical level, but the data has not been smoothed for any possible spatial correlation. (http://www.hqsc.govt.nz/assets/Health-Quality-Evaluation/Atlas/asthma-single/atlas.html. Date accessed: 2 July 2015).

The Pearson correlation coefficient is a commonly taught statistical technique in many statistical courses, and because it is familiar to many people, it may be tempting to use the statistic to assess agreement between methods, raters or instruments without understanding that correlation does not equate to agreement. The Pearson correlation coefficient measures the linear relationship between two continuous variables. For correlation and association studies, this is fine. However, when one is interested in examining agreement, it is incorrect to use the correlation coefficient, as it measures a different domain, and can give rise to different results. Statistics such as the intra-class correlation coefficient or the Bland–Altman plot would be more appropriate instead. As an example, suppose we are interested in assessing the level of agreement in quantifying the tumour volume size among 10 lung cancer patients between two radiologists A and B. As we can see from Figure 10.5, when the correlation coefficient is used, we get an excellent level of association ($\rho = 0.972$), which may be incorrectly construed by some as indicating excellent agreement instead. However, when we examine the mean values, it is clear that radiologist B consistently measures 5 units higher than radiologist A, which represents a 30% difference and can hardly be considered as a reasonable level of agreement, giving rise to a possible erroneous conclusion.

Similar to the correlation coefficient, the linear regression model is also popular in medical research and is used to examine linear relationships

pat_id	physiciana	physicianb
1	17	22
2	15	20
3	20	25
4	15	19
5	19	24
6	15	20
7	16	21
8	19	24
9	16	21
10	15	21

```
pwcorr physiciana physicianb
```

	physic~a physic~b
physiciana	1.0000
physicianb	0.9719 1.0000

```
su physiciana physicianb
```

Variable	Obs	Mean	Std. Dev.	Min	Max
physiciana	10	16.7	1.946507	15	20
physicianb	10	21.7	2.002776	19	25

FIGURE 10.5
Correlation and agreement.

between variables. Using the model for non-linear associations is problematic. For instance, suppose one is interested in the relationship between QoL as the outcome and body mass index (BMI) as the predictor variable. A linear regression model appears to be the correct model to use, especially because QoL is measured on a continuous scale and can be expected to follow a normal distribution. However, as we can observe from Figure 10.6, the relationship between QoL and BMI can hardly be characterised as linear. In fact, along with many other health outcomes, BMI has a curved relationship with QoL, with those who are underweight and obese showing a relatively poorer outcome compared to those who have a normal BMI range. Artificially fitting a straight line to a curve may not only give a bad fit to the data but result in erroneous results and conclusions, which can bypass many investigators' eyes especially if the data are not plotted and viewed. Solutions include categorising BMI in clinically meaningful groups (e.g. underweight, normal and obese) or fitting restricted cubic splines (Desquilbet & Mariotti, 2010).

Over-fitting is another possible issue not just in linear regression models, but also in other multi-variate modelling. This occurs when the model has a large number of parameters included (e.g. variables) in relation to the number of observations, and where the model fit is expected to be good based on the data that it is derived from. Model fit should be assessed externally (i.e. on prospective datasets or new patients) or the data could be split into two (derivation and validation samples), and other resampling techniques such as jackknife and bootstrap can also be used (Mooney & Duval, 1993).

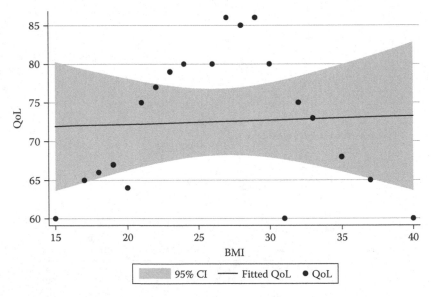

FIGURE 10.6
Plotting a linear relationship line when it does not exist.

Including additional covariates that may be co-linear is another potential problem when building multi-variate models. This may occur when the data analyst includes all the available variables in the dataset in the model without deliberating carefully about the usefulness of each variable or whether they are closely related to other variables already included in the model. In the earlier example, it is possible that the database includes height and weight variables that may be routinely collected to compute the BMI variable, which is the main covariate of interest in the study. Including weight or height alongside BMI in the multi-variate model will result in collinearity in the analysis, leading to incorrect inferences.

Failure to standardise data when comparing across groups or data across time points is another potential pitfall in the analysis of data. This is relatively more common in observational studies than in RCTs, where in the latter, any differences in the groups being studied can be attributed to chance due to the randomisation process. For example, in the health care setting, clinical quality indicators such as infection rates, re-admission rates and positive margin rates for prostate cancer surgery are usually compared across hospitals. This may be done in the form of a funnel plot, which consists of a scatterplot examining the indicators against a measure of precision (often the sample size). Control limits are usually set, such that they get narrower when the sample size increases, leading to the funnel shape of the plot. Data that lies outside of the contours of the funnel plot may possibly indicate much higher occurrence of the indicators for the hospital compared to the other hospitals, and may indicate possible systematic problems within the hospital that can be investigated. Figure 10.7 highlights funnel plots for hospital re-admissions following knee replacement surgeries among 16 hospitals. Figure 10.7a shows the crude plot, and it appears that a number of hospitals have re-admission rates that are much higher or lower than the rest of the hospitals. At first glance, it may appear that something may be wrong with the hospitals (i.e. patients not being optimally managed, etc.). However, after standardising for patient-level risk factors such as age, gender and post-surgery infections, Figure 10.7b now shows that most of the points fall within the expected level of re-admissions within the funnel plot. This indicates that most of the differences in re-admission rates across hospitals can instead be explained by demographic and clinical factors.

On a related note, sometimes the statistical analysis plan is not written with sufficient details to ensure reproducibility, which may not be a good idea. This can be seen in a manuscript, grant proposal or even an oral or poster presentation in a conference. For example, in Figure 10.8, we can see from this conference presentation slide that even though the models are stated (i.e. linear and logistic regression), it is unclear for which outcomes they were intended to be applied. In addition, it is not stated which of the tests, including *t*-test and ANOVA, relate to the corresponding means and groups to be compared in the study. From the details provided, it is hence not

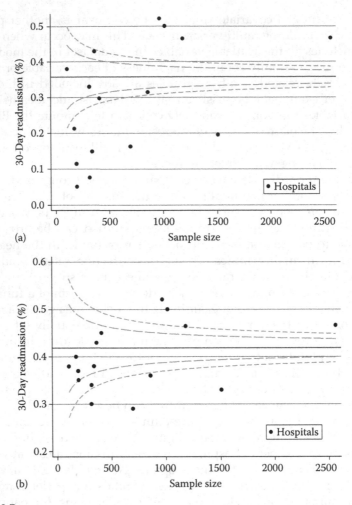

FIGURE 10.7
Funnel plots for hospital performance in terms of 30-day readmission following knee replacement surgery. (a) The crude plot and (b) the standardised funnel plot for patient-level risk factors such as age, gender and post-surgery infections.

possible to assess whether the correct statistical tests have been applied, and therefore difficult to ascertain the validity of the results of the study. When describing the analysis plan, it is important to mention the main outcome measure and the corresponding independent variables and the corresponding tests that are used. The groups that are being compared also need to be clearly stated. Altman (1991) has reported a study which found that only 46 of 132 controlled trials in cancer (35%) had specified the method of statistical analysis in the results, and this highlights the significance of the problem in the medical literature.

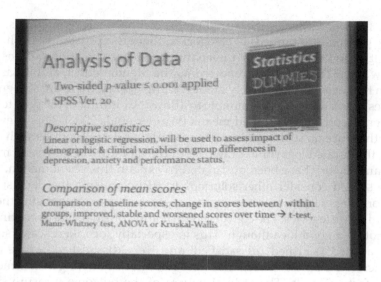

FIGURE 10.8
Insufficient details in a statistical analysis plan.

Sometimes, continuous outcome measures in a study are categorised or dichotomised for convenience in the analysis or ease of presentation of data in tables or charts. For instance, the continuous outcome variable may be severely skewed and transformation may not help, or one can expect a non-linear association for the explanatory variable under investigation. Other explanations such as 'Clinicians have to make dichotomous decisions to treat or not to treat, so it makes sense to have a binary outcome' or 'Physicians find it easier to understand the results when they're expressed as proportions or odds ratios' have also been offered for the decision to dichotomise (Streiner, 2002). Dichotomising data may lead to several problems, including the following (Altman & Royston, 2006):

1. Reduction in statistical power when looking for associations between variables.
2. Increase in the risk of positive result being classified as false positive.
3. Variation in outcome between groups may be underestimated, and considerable variability subsumed within groups.
4. Non-linearity in associations between variables may be masked.

For instance, it is popular practice to split the sample into normal/abnormal groups using cut-offs based on the distribution of the variable (e.g. >2 SDs from mean). The relative efficiency of this method has been shown to be a low value of 13% in terms of sample size planning for a satisfactory precision, and thus, this may not be a good strategy (Senn & Julious, 2009). This is also true for variables measured on an ordinal scale. Sometimes, when data is grouped (or binned) in categories, there is a possibility of showing correlation in the grouped data, even though such correlation was absent in the raw data. This has been reported as the Mendel effect, and it has been shown that even with relatively modest sample sizes, there are still enough degrees of freedom in a data sample to allow the binning of the data to yield apparent trends in the binned means (Wainer et al., 2006). This is even so when there are no such trends in the underlying data. Theories such as the law of the iterated logarithm drawn from the literature of Brownian motion and game theory have been postulated to explain this phenomenon.

One should consider other solutions to dichotomising the data, such as transforming the data in the event on non-normality or categorising the data into ordinal categories. The converse (i.e. treating ordinal data as continuous) may not be a good idea though. This is especially for Likert scales such as 'very satisfied', 'satisfied', 'dissatisfied' and 'very dissatisfied', where movement across consecutive categories does not infer a doubling in satisfaction/dissatisfaction levels. The ordinal nature of the outcome measure should then be preserved in the analysis using suitable statistical models, such as the ordinal logistic regression. When there is no choice but to categorise the data, clinically meaningful cut-offs should be used.

10.6 Not Checking for the Assumptions Behind the Test

According to Altman (1991), using methods of analysis when the assumptions are not met is one of the basic errors frequently made in journal articles. Ignoring the assumptions behind a statistical technique may lead to incorrect results and conclusions. Figure 10.9 shows the Stata output from a hypothetical dataset, comparing the mean years of smoking between groups of COPD and asthma patients. Results from the independent Student's t-test seem to indicate that asthma patients appear to have a significantly longer period of smoking compared to COPD patients ($p = .047$). However, one of the main assumptions of the test (i.e. normality of data) was violated and ignored (data not shown), and a subsequent analysis using the Mann–Whitney U test showed that there was indeed no significant difference between the two groups, with $p = .09$ (Figure 10.10). The data analyst should then ensure that he/she is familiar with all the assumptions behind a statistical test and that the assumptions are fulfilled.

Another misconception in the statistical analysis is that non-parametric tests are better than their parametric equivalents. For instance, parametric tests such the ANOVA test require certain assumptions such as normality of the data to be fulfilled, and these are often difficult to assess in studies with very small sample sizes. To get around the problem, some biostatisticians

```
. ttest yearsofsmoking, by( patients)

Two-sample t test with equal variances
```

Group	Obs	Mean	Std. Err.	Std. Dev.	[95% Conf. Interval]	
Asthma	20	30.55	2.999539	13.41435	24.27189	36.82811
COPD	20	22.7	2.372097	10.60834	17.73514	27.66486
combined	40	26.625	1.989294	12.5814	22.60127	30.64873
diff		7.85	3.824144		.1084248	15.59158

```
    diff = mean(Asthma) - mean(COPD)                         t =    2.0527
Ho: diff = 0                               degrees of freedom =        38

    Ha: diff < 0                 Ha: diff != 0                   Ha: diff > 0
 Pr(T < t) = 0.9765        Pr(|T| > |t|) = 0.0470           Pr(T > t) = 0.0235
```

FIGURE 10.9
Results from independent Student's *t*-test.

```
. ranksum yearsofsmoking, by( patients)

Two-sample Wilcoxon rank-sum (Mann-Whitney) test
```

patients	obs	rank sum	expected
Asthma	20	473.5	410
COPD	20	346.5	410
combined	40	820	820

```
unadjusted variance      1366.67
adjustment for ties        -3.33

adjusted variance        1363.33

Ho: yearso~g(patients==Asthma) = yearso~g(patients==COPD)
            z =    1.720
   Prob > |z| =    0.0855
```

FIGURE 10.10
Results from Mann–Whitney U test.

then automatically perform non-parametric tests without bothering to check for the assumptions behind the parametric equivalent of the test. The non-parametric tests are usually based on rank ordering and do not have parameters to characterise the distribution (hence the term non-parametric). They are also less efficient than parametric tests (i.e. require larger sample size to show statistical significance) with all else being equal and should be avoided unless assumptions for the parametric tests cannot be fulfilled. Non-parametric tests may be considered when there are extreme outliers in the data that are exercising undue influence on the results and cannot be removed or when the data are severely skewed that even transformation does not help.

The test for normality is a crucial assumption for many parametric tests such as the independent Student's *t*-test, linear regression model and ANOVA. To examine the distribution of a continuous variable, one can plot a histogram or even the quantile-normal plots. Formal tests such as the Shapiro–Wilk test can also be undertaken. The null hypothesis for the test is that the data comes from a normal distribution. However, one should be cautious about using such tests, especially for small samples. In some instances, the test may not have sufficient power to reject the null hypothesis even when there is obvious skewness in the data. However, for large sample sizes, the test will almost always suggest rejecting the null hypothesis that the data are normally distributed even with minor departures from normality. As we can see from Figure 10.11, even though the histogram indicates that the data are skewed, the Shapiro–Wilk test indicates that there is no evidence to reject the null hypothesis that the data appear to be normally distributed. This may lead to some data analysts choosing to perform parametric tests and possibly obtain erroneous results. In the event that data are skewed, one may wish to transform the data, as described earlier. However, some people may not like to work with transformed data, as the interpretation of the results may be a bit cumbersome. For instance, length-of-stay (LOS) data in the hospital is often positively skewed, and natural logarithmic transformation of the data is usually undertaken to establish normality. When one performs the linear

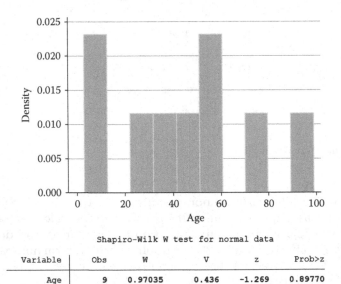

Variable	Obs	W	V	z	Prob>z
Age	9	0.97035	0.436	–1.269	0.89770

Shapiro–Wilk W test for normal data

FIGURE 10.11
Shapiro–Wilk test for normality.

regression analysis to study the factors associated with LOS, the interpretation of the coefficients from the final model can be a little difficult. With a bit of mathematical operation, it should not be too tedious to recover the interpretation of the coefficients on the original scale. For example, we have used the ratio of LOS between two departments, obtained from a log-transformed adjusted multi-variate regression model to work out the actual inter-departmental difference in LOS in a previous research article (Sahadevan et al., 2004).

10.7 Data Dredging

One habitual mistake by some investigators is to include all data available to them in the database (e.g. clinical, laboratory, administrative, etc.), and subsequently perform statistical modelling with all of these variables, hoping that some will achieve statistical significance out of chance alone. These variables are often selected without careful *a priori* reasoning or justification, and often lack a thorough review of the literature. One consequence of doing so is multiplicity and the resulting chance findings, which was initially discussed in Chapter 5. When multiple hypotheses are tested, one is likely to find some statistically significant results out of chance alone. The results may also not be reproducible by other researchers looking at the same problem. Sometimes, the results may not make sense clinically. As an example, a researcher performs a correlation analysis of the floor that his/her patients live on against their diabetes status and finds that those staying in particularly high floors 15–20 have a higher risk of diabetes than those staying in other floors! This is clearly non-sensical, unless these are highly subsidised rental units provided to the socio-economically disadvantaged elderly. There are a number of approaches that allow for the correction of multiple comparisons, including the family-wise error rate, Bonferroni correction and false discovery rate.

A possible sign of data dredging is the lack of *a priori* clearly stated statistical analysis plan along with aims and hypotheses. Failing to have a statistical analysis plan can give rise to a whole range of actions on the part of the investigators, which may lead to data dredging. This includes dichotomising or creating multiple categories out of an outcome measure, which was originally continuous, conducting various subgroup analysis, removing outliers without justification, including clinically non-important predictor variables that are highly significant and so on.

Data dredging should not be confused with serendipity or chance findings from the same experiment. For instance, Viagra was a potential treatment for angina and was being tested in clinical trials. As an angina treatment, it was pretty useless, but then the researchers began to get reports of some unexpected side effects. The other example is Fleming's discovery of penicillin, when a mould landed on his culture plate and killed off the bacteria. Also, data mining is an established field where large amounts of data are put through machine learning algorithms to produce insights and solutions, and this should not be mistaken for data dredging.

10.8 Presenting Tables and Figures Inappropriately

There are a number of points one can adhere to, to ensure that tables and figures created for manuscripts and presentations are appropriate. The KISS (keep it simple but scientific) principle should be adopted. A table that is cluttered with lots of information is not only difficult to read but does not captivate the reader. An example of a poster presentation that presents too much information in one table is shown in Figure 10.12. Chapter 7 covered some of the important points in creating tables and charts. Tables may be created within Microsoft Excel or customised software such as Latex. However, avoid pasting tables from the software output directly as these are often not of publication standard and unprofessional! (Figure 10.13).

When creating figures, one should avoid using colours (as journals often charge for colour printing) and use patterns instead to differentiate groups. The axis scales should also not be manipulated and must be standardised to avoid misinterpretation of the results. For example, as shown in Figure 10.14, it appears that there is a marked difference in the peak flow response between those on drugs A and B. However, from Figure 10.15, it appears that

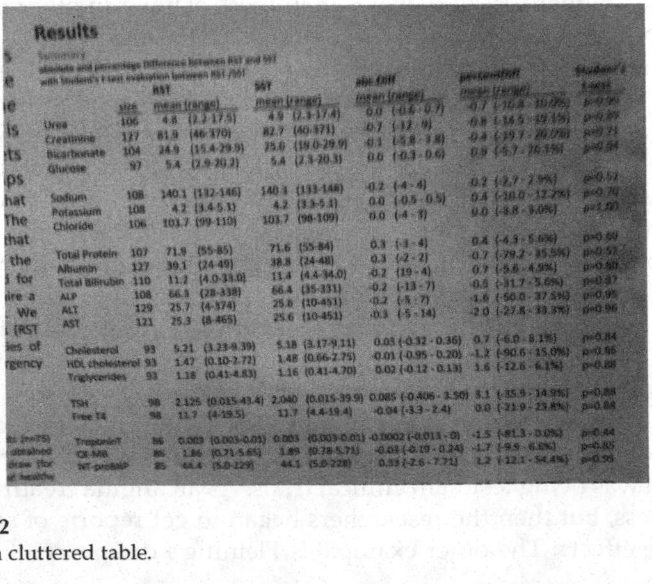

FIGURE 10.12
Example of a cluttered table.

Source	SS	df	MS		Number of obs =	576
					F(4, 571) =	9.82
Model	286820.628	4	71705.157		Prob > F =	0.0000
Residual	4169939.61	571	7302.87146		R-squared =	0.0644
					Adj R-squared =	0.0578
Total	4456760.23	575	7750.88736		Root MSE =	85.457

creatinine	Coef.	Std. Err.	t	P>|t|	[95% Conf.	Interval]
gender	-31.82587	7.134293	-4.46	0.000	-45.83853	-17.81321
ethnicity	13.64753	4.426386	3.08	0.002	4.953548	22.34152
age	1.006129	.2948358	3.41	0.001	.4270337	1.585224
hbaic	-.9346732	2.096229	-0.45	0.656	-5.051935	3.182588
_cons	81.84003	28.32898	2.89	0.004	26.19831	137.4817

FIGURE 10.13
Table from a Stata output.

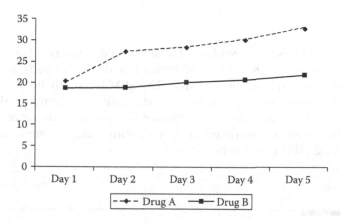

FIGURE 10.14
Peak flow with axis narrowed.

there is little difference between the two drugs when the axis values have been widened.

SDs are sometimes confused with the standard errors of the mean (SEMs). The SD describes the variability between individuals in a sample. The SEM describes the uncertainty of how the sample mean represents the population mean. Given that SEM = SD/square root(n), with n being the sample size; hence, the SEM is always lower than the SD. In a review of journal articles from four anaesthesia journals, Nagele (2003) found that the percentage of misuse of the SEM ranged from 11.5% in the *European Journal of Anaesthesiology* to 27.7% of articles in the *Anesthesia & Analgesia*. Reporting and reading the SEM incorrectly as the SD can have grave consequences. For instance, suppose a trialist wanted to plan for the sample size of a new two-arm drug trial looking at a mean change of 0.1 units, and incorrectly read the SEM

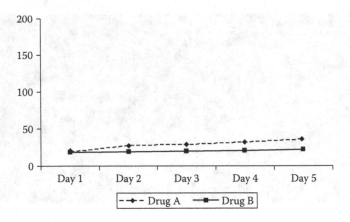

FIGURE 10.15
Peak flow with axis widened.

(instead of the SD) as 0.2 instead of 2. He/she would have ended up recruiting 64 patients per arm, instead of the 6280 that he/she really needs, resulting in the trial being severely under-powered. The SEM is usually used in calculating the confidence intervals, and in another related common mistake, sometimes when the 95% confidence intervals of the means in two groups do not overlap, the means are assumed to be not significantly different from each other, although this may not be true.

10.9 Reporting and Interpreting Data Inappropriately

One should also pay attention to careful interpretation of the results and avoid incorrect conclusions or possible 'spin' in the results. In a review of parallel-group RCT articles with a clearly identified primary outcome showing statistically non-significant results (Boutron et al., 2010), the authors found that 40% of the reports had spin in at least two of these sections in the main text. Spin was identified in the Results and Conclusions sections of the abstracts of 37.5% and 58.3% of the reports. The spin strategy consisted mainly of focussing on within-group comparison and subgroup analyses in the Results section. A 'spin' was defined as use of specific reporting strategies to highlight that the experimental treatment is beneficial, even though the results were not statistically significant. Clearly, this should be avoided as ineffective treatments or interventions would inadvertently end up being accepted in clinical practice.

p-Values are commonly not adequately reported in the medical literature or even misinterpreted. They should be presented to two significant digits; otherwise, it is difficult to ascertain statistical significance against the usual level of significance of 5% (e.g. $p = .047$ or $.053$). For really small *p*-values, computers usually round them off and present them as $p = .00000$. It is a bad idea to report this as $p = 0$, as *p*-values are never 0 or 1. Conventionally, they can be presented as $p < .001$. Sometimes, to save space in a table of results, asterisks or other symbols are given for statistically significant results (i.e. $p < .001$), and this information is denoted at the footnotes of the tables. However, one should not be overly-reliant on the presentation and interpretation of the *p*-values. Hubbard and Lindsay (2008) have published an article that presents several arguments on why the *p*-value is an inadequate measure of evidence in statistical significance testing, including *p*-values exaggerating the evidence against the null hypothesis and that they provide no information on the effect size. *p*-Values should be presented alongside 95% confidence intervals, which provide information on the effect size as well as variability. A wider confidence interval would infer greater variability and a smaller sample size, and vice versa. One general misinterpretation of the *p*-value is that it is the probability that the null hypothesis is true, so that a significant result means that the null hypothesis is very unlikely to be true (Sterne & Davey Smith, 2001).

When presenting results, authors need to also distinguish between statistical and clinical significance. For instance, describing the correlation between age and cholesterol ($\rho = 0.1$, $p < .001$) as being significant may be vague. This is because even though statistically the results appear significant, the effect size is weak and not clinically useful. Sometimes, there is also a tendency to interpret associations as being causative in nature. For instance, a significant relationship between coffee drinking and lung cancer may not necessarily mean that there is a compound in coffee that causes cancer, but more likely that coffee drinking is a possible confounder to another variable (smoking) that is predictive of development of lung cancer. Hill (1965) provides the following criteria for establishing causality: strength of association, consistency, specificity, temporality, biological gradient, plausibility, coherence, experimental evidence and analogy.

The other issue related to clinical significance lies in the presentation of regression coefficients for the explanatory variables under investigation. For example, rather than reporting that LOS increases by 0.1 day for every 1 mg/dL increase in total cholesterol, it may be clinically more meaningful to report that LOS increases by 1 day for every 10 mg/dL increase in cholesterol. This re-scaling of continuous explanatory variables also includes using the interquartile range as a scaling factor, so that interpretation is intuitive in that we are comparing those in the middle of the upper part of the cholesterol distribution with those in the middle of the lower portion of the distribution.

Another potential mistake in the interpretation of trial results is to confuse 'evidence of absence of effectiveness' with 'absence of evidence of effectiveness'. The latter is usually performed in a study to determine effectiveness between two drugs A and B. When the results are not significant, this is sometimes incorrectly interpreted as drugs A and B being equivalent (i.e. evidence of absence of effectiveness). The set-up of the basic hypothesis is to be tested; sample size consideration and analytical methods between the two approaches are very different.

In a closely related topic, one should ensure that all the outcomes initially planned in a study protocol and proposal are reported at the end of the study. This is to ensure that the published studies do not provide a biased estimate of the effect being studied and to ensure transparency. This appears to be a prevalent problem, especially reported for RCTs. For instance, starting from protocols and protocol amendments for randomised trials approved by the Scientific-Ethical Committees for Copenhagen and Frederiksberg, Denmark, over a period of 2 years, one study then went on to examine the final published results and looked for selective reporting of outcomes (Chan et al., 2004). It was reported that 92% of the trials had at least one incompletely reported efficacy outcome, whereas 81% had at least one incompletely reported harm outcome. Major discrepancies between the primary outcomes specified in protocols and those defined in the published articles were found in 62% of the articles. To prevent such practices, study protocols should be made available to the public (e.g. published online) at the earliest convenient time before the start of the study. Deviations to the study protocol should also be described in the manuscript.

When one interprets the data that has been presented, it is worthwhile to note whether the study results could have been unduly influenced by external parties (e.g. sponsorship of study and/or financial support of principal investigator from external private companies). For example, in a review of English medical journal articles from March 1995 to September 1996, looking at the controversy about the safety of calcium-channel antagonists, the study classified articles as being supportive, neutral, or critical with respect to the use of calcium-channel antagonists. The authors' financial relationships with both manufacturers of calcium-channel antagonists and manufacturers of competing products were also obtained via a survey questionnaire (Stelfox et al., 1998). The study had a good response rate of 80% and reported that 96% of the supportive authors had financial relationships with manufacturers of calcium-channel antagonists, compared with 60% of the neutral authors and 37% of critical authors. The association was also found to cut across the various categories of financial support, including funds for travel expenses, honorariums for speeches, support for educational programs, research grants, and employment or consultation. As an author or collaborator in an industry-funded project, one should ensure that all conflicts of interests are declared, and that the funding authority does not have a role in the design, analysis and interpretation of the study.

10.10 Conclusion

In sum, it is probably unwise to ignore the dangers of incorrect design, analysis and reporting of a study. One needs to be aware of the following common pitfalls:

1. Choosing the inappropriate design or sub-optimal features within a study design
2. Selecting too few subjects for the study and not accounting for missing observations
3. Poor understanding of randomisation and its features
4. Incorrect application of statistical tests
5. Ignoring key assumptions behind a statistical test
6. Poor presentation of tables and figures
7. Misinterpretation of data and/or providing spins in results

Key Learning Points

1. Identify the common mistakes made in the design of a study, including sample size calculations.
2. Learn how to avoid choosing inappropriate statistical tools to analyse your data.
3. Recognise the implications of poor analytical practices such as data dredging and failure to check for assumptions behind statistical techniques used.
4. Gain useful tips on how not to present your data.
5. Avoid misinterpreting data and providing 'spin' in the writing of the results.

References

Altman, D. G. (1991). *Practical Statistics for Medical Research*. London: Chapman and Hall/CRC Press.

Altman, D. G., & Royston, P. (2006). The cost of dichotomising continuous variables. *BMJ, 332*(7549), 1080. doi:10.1136/bmj.332.7549.1080.

Boutron, I., Dutton, S., Ravaud, P., & Altman, D. G. (2010). Reporting and interpretation of randomized controlled trials with statistically nonsignificant results for primary outcomes. *JAMA, 303*(20), 2058–2064. doi:10.1001/jama.2010.651.

Carlin, B. P., & Louis, T. A. (2000). *Bayes and Empirical Bayes Methods for Data Analysis.* New York: Chapman & Hall/CRC Press.

Chan, A. W., Hrobjartsson, A., Haahr, M. T., Gotzsche, P. C., & Altman, D. G. (2004). Empirical evidence for selective reporting of outcomes in randomized trials: Comparison of protocols to published articles. *JAMA, 291*(20), 2457–2465. doi:10.1001/jama.291.20.2457.

Desquilbet, L., & Mariotti, F. (2010). Dose-response analyses using restricted cubic spline functions in public health research. *Stat Med, 29*(9), 1037–1057. doi:10.1002/sim.3841.

Earnest, A., Morgan, G., Mengersen, K., Ryan, L., Summerhayes, R., & Beard, J. (2007). Evaluating the effect of neighbourhood weight matrices on smoothing properties of Conditional Autoregressive (CAR) models. *Int J Health Geogr, 6*, 54. doi:10.1186/1476-072X-6-54.

Earnest, A., Tan, S. B., Wilder-Smith, A., & Machin, D. (2012). Comparing statistical models to predict dengue fever notifications. *Comput Math Methods Med, 2012*, 758674. doi:10.1155/2012/758674.

Egger, M., Davey Smith, G., Schneider, M., & Minder, C. (1997). Bias in meta-analysis detected by a simple, graphical test. *BMJ, 315*(7109), 629–634.

Harris, A. H. S., Reeder, R., & Hyun, J. K. (2009). Common statistical and research design problems in manuscripts submitted to high-impact psychiatry journals: What editors and reviewers want authors to know. *J Psychiatr Res, 43*, 1231–1234.

Hill, A. B. (1965). The environment and disease: Association or causation? Proceedings of the Royal Society of Medicine, vol. 58, 295–300.

Hubbard, R. & Lindsay, R. M. (2008). Why p values are not a useful measure of evidence in statistical significance testing. *Theor. Psychol., 18*(1), 68–88.

Keen, H. I., Pile, K., & Hill, C. L. (2005). The prevalence of underpowered randomized clinical trials in rheumatology. *J Rheumatol, 32*(11), 2083–2088.

Mooney, C. Z., & Duval, R. D. (1993). *Bootstrapping: A Nonparametric Approach to Statistical Inference.* Newbury Park, CA: Sage.

Nagele, P. (2003). Misuse of standard error of the mean (SEM) when reporting variability of a sample: A critical evaluation of four anaesthesia journals. *Br J Anaesth, 90*(4), 514–516.

Sahadevan, S., Earnest, A., Koh, Y. L., Lee, K. M., Soh, C. H., & Ding, Y. Y. (2004). Improving the diagnosis related grouping model's ability to explain length of stay of elderly medical inpatients by incorporating function-linked variables. *Ann Acad Med Singapore, 33*(5), 614–622.

Senn, S., & Julious, S. (2009). Measurement in clinical trials: A neglected issue for statisticians? *Stat Med, 28*(26), 3189–3209. doi:10.1002/sim.3603.

Stelfox, H. T., Chua, G., O'Rourke, K., & Detsky, A. S. (1998). Conflict of interest in the debate over calcium-channel antagonists. *N Engl J Med, 338*(2), 101–106. doi:10.1056/NEJM199801083380206.

Sterne, J. A. & Davey Smith, G. (2001). Sifting the evidence-what's wrong with significance tests?. *BMJ, 322*(7280), 226–231.

Streiner, D. L. (2002). Breaking up is hard to do: The heartbreak of dichotomizing continuous data. *Can J Psychiatry, 47*(3), 262–266.

Wainer, H., Gessaroli, M., & Verdi, M. (2006). Finding what is not there through the unfortunate binning of results: The Mendel effect. *Chance, 19*(1), 49–52.

11

Views from the Ground: A Survey among Biostatisticians and a Chat with Clinicians

11.1 Introduction

Currently, the collaborative process and experience between biostatisticians and their collaborators has not been reported collectively, and little is known about the challenges faced and opportunities available to improve the process. A survey was conducted among members of the Centre for Quantitative Medicine (CQM) at the Duke-NUS Graduate Medical School in Singapore. The purpose of the survey was to obtain a baseline understanding of the experiences of statisticians specifically in terms of collaborating with clinicians and other researchers. This is the first time such a survey has been conducted and reported. Section 11.2 in this chapter provides details on the survey methodology. Topics studied in the survey such as problems faced in collaborating with clinicians, sources of training on collaboration skills, common responsibilities of an academic biostatistician and enhancing collaborations and skills to gain to further collaborations are subsequently presented. Finally, interviews with selected clinicians from diverse medical specialties and geographical regions, such as the United States, the United Kingdom, Singapore and Australia, were conducted to better understand the collaborative process from the other end of the collaboration spectrum.

11.2 Survey Methodology and Profile of Respondents

The brief 10-item self-administered survey (see the appendix for a copy of the questionnaire) was conducted in July–August 2013 using the SurveyMonkey online software (https://www.surveymonkey.com. Date accessed: 24 June 2015). Prior to administration of the survey, the questionnaire was pre-tested and reviewed for content validity by a fellow faculty member who was experienced in psychometric scales and qualitative analysis. The target audience was the CQM that is an academic home consisting of quantitative scientists, which strives to bring biomedical research and the quantitative science communities together. The CQM includes academic faculty from the Duke-NUS Graduate Medical School in Singapore, which hosts the Centre as well as other members who come from a diverse range of research institutes, including Singapore General Hospital, National Cancer Centre, Singapore Eye Research Institute, National Healthcare Group, Singapore Clinical Research Institute, National Heart Centre and Genome Institute of Singapore, most of which are all geographically located within the same campus.

A link to the online survey was sent out to all 47 members of CQM via email. A follow-up reminder was sent 2 weeks before the deadline. A total of 32 members replied, yielding a response rate of 68%. In terms of the highest level of educational qualification, 13% had a PhD, 47% a master's degree and the rest 40% held a bachelor's degree. The majority, 63%, came from an academic background, whereas the rest 37% were from a hospital/health care institution setting. The median years of working as a biostatistician/data analyst was 4 (interquartile range [IQR] 2–10). In the past 12 months, the median number of collaborative projects involved as a biostatistician/analyst was 7 (IQR 3–12).

11.3 Problems Ever Faced in Collaborating with Clinicians

The top three problems (Table 11.1) faced by biostatisticians when collaborating with clinicians were as follows: (1) Collaborator takes a long time to write up a manuscript or grant proposal (55%), (2) collaborator approaches a biostatistician at the last minute (42%) and (3) a tie between difficulty in explaining statistical concepts and collaborator engages in data dredging (39%). Other comments made include 'no respect and appreciation shown to

TABLE 11.1

Problems Ever Faced in the Course of Collaborating with Clinicians/Health Researchers

Problems	n (%)
Difficulty in understanding medical terms	11 (35)
Difficulty in explaining statistical concepts	12 (39)
Unrealistic deadlines set by collaborator	6 (19)
Collaborator approaches me at the last minute	13 (42)
Collaborator engages in data dredging	12 (39)
Authorship issues	8 (26)
Collaborator takes a long time to write up manuscript/grant proposal	17 (55)

a biostatistician…' and 'some collaborators do not consider the statistician as a scientist but only an analyst….'

The first problem of the collaborator taking too long to write up the manuscript is a real one, especially in the hospital setting where for many practicing clinicians, doing research may be more of a hobby and done in their personal time, often during weekends.

Depending on the emphasis the clinician's institution places on research, the amount of time the clinician has for research varies. Other factors are equally important, such as the clinician's interest and research experience. When faced with such a situation, there are a couple of things the collaborating biostatistician can do. Firstly, regular reminders can be sent to the busy clinician, enquiring about updates on the manuscript. Often, it helps to ask if there is anything else the biostatistician needs to do to help with the completion and submission of manuscript (e.g. additional analysis, writing the results or statistical sections for the manuscript, etc.). It also pays to discuss the anticipated manuscript submission date with the clinician during the first meeting, so that expectations can be managed and a deadline is available to work towards. The use of a project file for this purpose was discussed in Chapter 8. Sometimes, the same clinicians may have a number of concurrent projects with the biostatistician, and it may help to prioritise and work with the clinician on one project at a time and only start on the next project once the first project is completed or submitted for publication/grant application.

'Collaborators approaching the biostatistician at the last minute' is a common problem. One possible reason is that demand for biostatistical services in the hospital setting is often seasonal, with demand picking up when dates for the submission for grant application approach close to institutional deadlines, during medical conference events, and when Advanced Specialty Training (AST) students expect to graduate (often needing to publish papers in the process). Some collaborators could also have an unrealistic expectation on the timeline for biostatisticians to revert with the analysis, without realising that there may be several concurrent projects the biostatistician is working on. In the words of a senior colleague of mine, 'What must he/she be thinking? That I am sitting down and doing nothing, waiting for his/her data to arrive….' Every project is an important and urgent project for the collaborator, and it is important that the biostatistician manages his/her expectation early during the collaboration.

It is also important that institutions or departments set policies in place on how collaborators should engage with a biostatistician. For instance, there should be a minimum period (say at least 3 months) prior to a grant submission date before the biostatistician should be approached and expected to agree to the collaboration. A similar period can also be suggested for a collaboration leading to a publication. However, the actual time required for the project would depend on the stage of the project (e.g. planning, data collection, complexity of analysis, etc.) and the research experience of the collaborator. Therefore, the key element is to scope out the nature of the project during the first meeting with the collaborator and to discuss and agree on the expected timelines for data analysis, manuscript writing and submission before the start of the collaboration. Chapter 8 discussed the use of standard operating procedures (SOPs) to set a framework on engaging biostatisticians for collaborative research. In reality though, sometimes, it is difficult for the biostatistician to say no to last-minute projects, particularly if the request comes from someone senior in the organisation (e.g. head of school or the boss!). Projects may also move up/down the priority list of the biostatistician due to a number of reasons, including principal investigator may be too busy to work on project or the outbreak of a new infectious disease that may necessitate the top priority.

Chapter 9 provided strategies to deal with collaborators who request for unreasonable turnaround time for projects. It helps for the biostatistician to keep a tab of projects that are being worked on, as well as the time spent on each project so that this information can be provided as a justification to the institution to employ more biostatisticians if necessary. Getting the collaborator to show respect and treat the statistician as an equal is a bit tricky. The biostatistics profession is relatively new in the medical field, and many clinicians may not have any prior experience in working with biostatisticians. Over time, this should get better, particularly when the biostatistician can add value to the projects and win the respect of the clinicians. Spending enough time to discuss the clinical problem, making a name for oneself in the statistical field and fostering good working relationship with the collaborator are some steps biostatisticians can take to help develop a stronger sense of respect in the working relationship. In Chapters 8 and 9, we learnt some of these strategies in working with our collaborators.

11.4 Training on Collaboration/Consultation Skills

In terms of sources of training, the top three were (1) learning from other colleagues (78%), (2) learning from collaborators themselves (63%) and (3) formal school course (41%). Books/journals, structured on-job training and websites/other electronic resources were less likely to be sources of training (Table 11.2). Other sources mentioned included 'own experience and training as a graduate student....' In the United Kingdom, postgraduate training involves a 1-year full-time MSc in medical statistics or a PhD course (see, e.g. the Department of medical statistics at the London School of Hygiene & Tropical Medicine. http://www.lshtm.ac.uk/eph/msd/index.html. Date accessed: 19 June 2015). In the United States, the profession is known as biostatistics, and the MSc program is usually 2 years and based on a mix of credit-based courses and a thesis (e.g. Duke department of biostatistics

TABLE 11.2

Source of Training on Collaboration/Consultation Skills

Source	n (%)
Formal school course	13 (41)
Books/journals	10 (31)
Learning from other colleagues	25 (78)
Structured on-job training	7 (22)
Websites and other electronic resources	12 (38)
From collaborators themselves	20 (63)

and bioinformatics. http://biostat.duke.edu/master-biostatistics-program/curriculum. Date accessed: 19 June 2015). In Australia, a master's degree in biostatistics consisting of 12 units of study, including a Workplace Project Portfolio offered by the Biostatistics Collaboration of Australia (BCA) (http://www.bca.edu.au/courseinfo.html. Date accessed: 19 June 2015). The BCA is a consortium of biostatistical experts from around Australia with representatives from universities, government and the pharmaceutical industry. Although most programs have a project-based thesis that is often centred on a real-life clinical problem and may provide interactions with clinicians, training on collaboration/consulting skills is not a mainstream course.

Slightly less than half (45%) of the respondents said that they were very/completely satisfied with their current skills in collaborating/consulting with clinicians/health researchers. The factors associated with satisfaction with their current skills were explored using a binary logistic regression model. The explanatory variables included educational level, type of institution employed in, years of working experience, number of collaborative projects involved in the past year and the various sources of training obtained. Even with a small sample size as a limitation in this analysis, two important factors emerged as significant: (1) Those who had obtained training from their collaborators were 6.8 times more likely to be satisfied with their current skills (95% confidence interval [CI]: 1.1–39.8) than those who did not ($p = .035$) and (2) the odds ratio (OR) of satisfaction increased by a factor of 2.4 for every 10 increase in collaborative projects involved in the past year, but the results were marginally significant ($p = .097$). It is interesting to note that working directly with their collaborators offers greater satisfaction on their current collaboration skill levels than learning from colleagues. The experienced collaborator is a rich source of information, as often he/she is not only knowledgeable about the clinical information on projects, but has probably amassed enough information on research methodology and statistics revolving around his/her clinical specialty. For instance, a radiologist would be more familiar with statistical methods of agreement such as kappa than a cardiologist, and this could be through reading related articles and interacting with experienced colleagues and other statisticians. In a larger and more established statistical unit, it may be good practice to assign junior statisticians to specific medical specialties so that they can learn from their collaborators and gain experience over time. The finding that biostatisticians get more satisfied with their collaboration skills with an increase in collaborative projects is hardly surprising but a compelling reason to expose junior statisticians to more collaborative work.

11.5 Frequency of Performing Selected Tasks as a Biostatistician

Table 11.3 shows the tasks performed by a collaborative biostatistician in a typical month. The top three frequently performed tasks (often and always) were (1) data analysis (97%), (2) data management (70%) and (3) statistical analysis plan (59%). Other comments provided by respondents included 'manuscript writing, reviewing, educating on best practices, self-studying subject area being analysed, participating in study conceptualization and study design, reading on a new method that may be more appropriate, secondary analyses,...'

Although it is not unexpected to find that data analysis is the most frequently performed task for a biostatistician, it is surprising to find that data management forms the next most frequent task. Although it is fine to undertake some of the data management tasks where appropriate (e.g. recoding and generating variables and labelling of variables), undertaking substantial data management work like data entry, transcribing text data and inputting and merging data from non-traditional sources like a word processing file should be avoided. These may not be a cost-effective task for a biostatistician to undertake and are more appropriately handled by other research staff such as a clinical research coordinator or a research assistant.

TABLE 11.3

Frequency of Tasks Performed during Collaborations in a Typical Month

Tasks	Never/Almost Never	Seldom	Often	Always
Data analysis	1 (3%)	0 (0%)	17 (55%)	13 (42%)
Statistical analysis plan	1 (3%)	12 (39%)	16 (52%)	2 (7%)
Sample size/power calculation	4 (13%)	10 (33%)	13 (43%)	3 (10%)
Hypothesis formulation	3 (10%)	12 (39%)	11 (35%)	5 (16%)
Data management	1 (3%)	8 (27%)	14 (47%)	7 (23%)
Randomisation	10 (33%)	16 (53%)	3 (10%)	1 (3%)
Questionnaire/form design	13 (45%)	9 (31%)	6 (21%)	1 (3%)

Next, factors associated with performing data management tasks frequently were examined. Those who indicated that they did not obtain any form of training on collaboration/consultation skills (whether from formal courses, books, colleagues, etc.) were 4 times more likely to perform data management tasks regularly, although the result was not statistically significant, probably due to small numbers ($p = .101$). This is hardly surprising as clinicians and collaborators value statisticians who have obtained some form of training in collaboration skills and able to work on problems in applied medical research. The onus is on the practicing statistician to continually upgrade his/her skills in collaborating with others. Interestingly, we also found that those who ranked 'understanding medical terminology' least important as a skill to gain were significantly more likely to perform data management tasks (OR = 2.24), with 95% CI: 1.14–4.39, $p = .019$. This seems to indicate that those who were not exposed to medical terminology (i.e. not experienced in medical research) end up in a data management role. Another possible reason could be that the statistician often undertakes a passive role in the collaboration and ends up collecting a large number of variables in the dataset for analysis, rather than understand the nature of the project and question and limit the inclusion of variables in the study by discussing with the clinician key variables of interest.

11.6 Issues to Address in Order to Enhance Greater Collaborations

The survey revealed that the top three issues (Table 11.4) to resolve in order to enhance greater collaborations were (1) communication between biostatisticians and clinicians (78%), (2) technical knowledge of biostatisticians (59%) and (3) providing better matching with specific biostatistics skill that may be required for project (59%).

TABLE 11.4

Issues to Address in Order to Enhance Greater Collaborations with Clinicians

Issues	*n* (%)
Communication between biostatisticians and clinicians	25 (78)
Accessibility and availability of biostatisticians	12 (38)
Technical knowledge of biostatisticians	19 (59)
Increasing future pool of trained biostatisticians	18 (56)
Providing better matching with specific biostatistics skill that may be required for project	19 (59)
Providing institutional requirement that all clinical research projects should include a biostatistician collaborator	18 (56)

Other suggestions included 'target the right clinicians because research is not for all, improve quality rather than quantity of collaborations and providing a team of two methodologist, a senior and a junior, to handle a request. This provides opportunity to learn stats handling, collaboration tactics, catching errors and back up support....'

An important result from this survey has been the identification of 'Communication between biostatisticians and clinicians' as an issue that can enhance collaboration with clinicians. Indeed, clinicians may find it difficult to understand statistical terminology and are not used to mathematical notations and symbols, mainly because they may not have been exposed to it in medical school. Similarly, statisticians may not have exposure to medical jargons, abbreviations or epidemiological or research methodological terms, unless they have taken a postgraduate degree in biostatistics or equivalent research methodology courses previously or worked in a medical research setting. This is such an important topic that an entire Chapter 6 has been dedicated to it.

In order to better match projects between statisticians and collaborators, it is important to have an updated database on specific skill sets of biostatisticians as well as details on projects that need collaborative input, including aims/hypotheses and study design. In addition, it would be good to have an administrative person (preferably the head of the statistics department or someone senior in the department) to assign suitable collaborative projects to statisticians. It is equally important to follow up closely on the collaboration to ensure that the project is going smoothly. Sometimes, collaborations may not proceed smoothly because the two personalities are not getting along well, and if mediation efforts fail, it may be necessary to assign someone else to the project instead. Senior biostatisticians should also volunteer their time in institutional ethics boards, journal editorial boards and even grant review panels to help assess the statistical quality of manuscript and grant submissions, and hopefully suggest the need for biostatistical collaborations in those projects to raise the level of statistical literacy.

11.7 Skills Most Important to Gain to Improve on Collaborations with Clinicians

Table 11.5 highlights the skills that were identified by the respondents to be important in terms of improving collaborations with clinicians. The top three skills were (1) advanced statistical models (45% ranked most important), (2) communication skills (29%) and (3) project management and understanding medical terminology (10%).

TABLE 11.5

Skills Most Important to Gain to Improve on Collaborations with Clinicians

Skills	1 (Most Important)	2	3	4	5 (Least Important)
Advanced statistical models	14 (45%)	9 (29%)	0 (0%)	4 (13%)	4 (13%)
Time management	2 (6%)	6 (19%)	11 (35%)	7 (23%)	5 (16%)
Project management	3 (10%)	8 (26%)	9 (29%)	6 (19%)	5 (16%)
Communication skills	9 (29%)	4 (13%)	6 (19%)	8 (26%)	4 (13%)
Understanding medical terminology	3 (10%)	4 (13%)	5 (16%)	6 (19%)	13 (42%)

Courses on advanced statistical models such as Bayesian analysis or longitudinal modelling may be offered by universities as part of coursework for a postgraduate course or even short courses (some examples are provided later, but these are not exhaustive). Topics may also include missing data imputation technique, multi-level modelling, geo-spatial modelling and time series analysis. Just like other fields in medicine, the biostatistical literature is always evolving. Hence, it is important for biostatisticians to continually upgrade their skills by attending courses and seminars, statistical or medical conferences and reading the latest books in the related fields (e.g. Chapman & Hall/CRC Biostatistics Series).

11.7.1 Formal Postgraduate Degrees

1. Masters in medical statistics. London School of Hygiene and Tropical Medicine. http://www.lshtm.ac.uk/study/masters/msms.html. Date accessed: 22 January 2015.

2. PhD in biostatistics. Duke department of biostatistics and bioinformatics. http://biostat.duke.edu/phd-program. Date accessed: 22 January 2015.

3. Masters degree and graduate diploma in biostatistics. Biostatistics Collaboration of Australia. http://www.bca.edu.au/. Date accessed: 22 January 2015.

11.7.2 Short Courses

1. Graduate Summer Institute of Epidemiology and Biostatistics. John Hopkins School of Public Health. http://www.jhsph.edu/departments/epidemiology/continuing-education/graduate-summer-institute-of-epidemiology-and-biostatistics/. Date accessed: 22 January 2015.

2. Biostatistics for Clinical and Public Health Research. Monash University. Biostatistics for Clinical and Public Health Research. http://monash.edu/news/notices/show/biostatistics-for-clinical-and-public-health-research-3. Date accessed: 22 January 2015.

3. Biostatistics courses. Harvard School of Public Health. http://www.hsph.harvard.edu/biostats/courses/descriptions/courses_1415.html#summer. Date accessed: 22 January 2015.

11.7.3 Online Courses

1. Introduction to biostatistics. Yale School of Medicine. http://ycci.yale.edu/education/onlinecourses/biostatistics.aspx. Date accessed: 22 January 2015.

2. Advanced course in epidemiological analysis. London School of Hygiene and Tropical Medicine. http://www.lshtm.ac.uk/study/cpd/sacea.html. Date accessed: 22 January 2015.

11.7.4 Journals (Development and Application of Statistics in Medicine)

1. *Statistics in Medicine.* http://onlinelibrary.wiley.com/journal/10.1002/(ISSN)1097-0258. Date accessed: 22 January 2015.

2. *Biometrics.* http://onlinelibrary.wiley.com/journal/10.1111/(ISSN)1541-0420. Date accessed: 22 January 2015.

3. *Statistical Methods in Medical Research.* http://smm.sagepub.com/. Date accessed: 22 January 2015.

11.8 Views from Clinicians Who Have Collaborated with Biostatisticians

In this section, interviews with four clinicians who are experienced in clinical research and come from a variety of clinical specialties, including rheumatology and immunology, infectious diseases and psychiatry were conducted. The aim is not to get a representative view among all clinicians, but rather to understand the biostatistics collaboration from the perspective of the clinician through a qualitative process. In particular, views were sought on how a biostatistician has added value to their projects, challenges in collaborating with the statistician and skills and resources that should be made available to the statistician. There were also some valuable suggestions made on how to improve the relationship between the statistician and the clinician in a collaborative process.

11.8.1 Interview with Dr. Leong Khai Pang, MBBS, FRCPE, FAMS

Senior consultant and adjunct associate professor
Department of rheumatology, allergy and immunology
Assistant chairman (Medical Board)
Director, Clinical Research Unit
Tan Tock Seng Hospital
Singapore
11 March 2014

1. **In the past 5 years, how many biostatisticians or data analysts have you worked with?**

   I have worked with about seven to eight biostatisticians. This ranges from consulting them for simple statistics to collaborating with them on more engaging projects.

2. **In what ways have biostatisticians added value to your research projects?**

   Biostatisticians provide input in the correct design, analysis and interpretation of study results. Sometimes, they also help to check that the analysis done by a clinician or other researcher is correct, or even to make sure that the assumptions behind each test has been fulfilled.

3. **Can you describe some of the challenges you may have faced in collaborating with biostatisticians?**

   First, it is their background. If they have no biology or medical background, it makes it a bit more difficult. Also, they need to understand what goes on in the clinic, so as to understand the research question a little better. There is a lack of innate knowledge about how a doctor treats a patient, or the biological process behind a disease. For example, when doctors or nurses see patients, it is not a one-off thing, but rather part of regular followup sessions the patients have with the health care provider. This medical knowledge is something that statisticians can gain through regular interactions with the clinicians, either from on-job training or by visiting the clinics. Because of the nature of their work, it is not conducive for doctors and nurses to do research during office hours. In order to make themselves accessible to researchers, the statistician may need to carry out their consultation after office hours and work on the projects during office hours. Clinicians face a problem with deadlines in research; for instance, there is often an urgency to publish within a short period of time. Some statisticians need to face up to this challenge and tell their collaborating clinician that their data are not ready, making too many assumptions or dropping variables/observations without any justification, for instance. They need to stand up and say what is right. However, there needs to be a balance. You cannot have a statistician who keeps challenging the clinicians. Each query needs to be backed up with the facts and figures.

4. **Are there any specific knowledge or skill sets that biostatisticians should acquire to raise the standard of collaborative input?**

Project management is important, so that they understand that research projects have timelines, time management and communication skills. These should be taught in schools if they are not already. Spoken communication in particular is important so that the statistician does not come across as pushy: it is all about gaining trust with the clinician. Writing skills are also important. For instance, the statistician should write out the statistical analysis plan section and parts of results portion of protocols or manuscripts. Statisticians should also play a bigger part in writing grants, be it their own or supporting clinicians to write one, specifically the power calculations and study design sections in addition to the statistics.

5. **What are the resources available to statistician to gain these skills you've mentioned?**

There are books and courses available for non-clinicians to understand medicine. For example, medical typists in the past and clinical coders have used these resources. Perhaps these should be made available to statisticians as well. Regular interactions with clinicians are also important, as with mentorship from more experienced senior statisticians. I have met many young statisticians and epidemiologists who have claimed to have studied logistic regression in class before, but have not applied it to real-life data before. The reality is that classroom data are clean and pure and detailed! Whereas actual clinic data are often messy with missing data. Therefore, the statistician needs to be given real-life dataset to train on.

11.8.2 Interview with Professor Nick Paton, MB BChir (Cambridge), MRCP (Internal Medicine) (London), MD (Cambridge), FRCP(Edinburgh), DTM & H (London)

Research director
University Medicine Cluster
National University Hospital
Singapore
12 March 2014

1. In the past 5 years, how many biostatisticians or data analysts have you worked with?

That would be around six to seven biostatisticians, including three to four in the Medical Research Council in the United Kingdom.

2. In what ways have biostatisticians added value to your clinical trials?

The best statisticians I have worked with are involved in the concept stage of a trial. They have extensive therapeutic area expertise, and know the ins and outs of human immunodeficiency virus and tuberculosis, which are my main areas of work in the past 5 years or so. Statisticians contribute very early on to the research questions, then going onto the trial study designs, such as selecting the optimal study end points, and discussing issues around duration of trial followup, monitoring strategies, sample size and statistics, and so on. In recent years, I have worked on grant applications, often with the statistician as the main other applicant. In fact, as a reviewer, I often look at the application's team composition for a statistician, because if there is not, that is going to be pretty strong signal that the trial is going to be a disaster. If the principal investigator or other scientists do not have a good working relationship with an experienced statistician, then important things are going to be missed in the trial, including data management. The best guarantor for a good trial is to see if a statistician has been included as a main applicant. This is just in the planning stage of a trial. Obviously, in the implementation stage, the statistician plays a pivotal role, including whether case report forms are properly created, appropriate data capture system has been adopted and so on. A senior statistician is likely to have worked on a number of clinical trials in one or a small group of therapeutic areas, so they would have good insights into particular tools such as outcome measures, and also bring experience onto table on trial management, based on what works and what does not work, from previous studies they had been on.

3. **Can you describe some of the challenges you may have faced in collaborating with biostatisticians?**

As far as I can remember, there have been no bad collaborations! The problem is that some people may not be clear about what they want, or a junior investigator may expect too much knowledge about the therapeutic area from a statistician.

If you know what you want out of collaboration, and if this is expressed very clearly to a statistician, it does not become a challenge anymore. Perhaps 15 years ago, when I collaborated with a biostatistician, conversations were more tenuous and frustrating, maybe because I did not know what I wanted from the statistician, and vice versa. Therefore, when an investigator starts out his/her research career, there can be some challenges in collaborating with biostatisticians in terms of expectations from each side and the type of expertise each one brings to the table.

4. **Have you faced challenges specifically related to turnaround time from statisticians?**

I think it is all down to good scientific methodology. Ideally, you sort out at the start of the project exactly what the question is, exactly how you are going to approach the analysis to answer the question, make sure that your data collection is done the right way, so that when the data collection is done, the analysis is relatively straightforward. The problem arises when these issues are not sorted out in the beginning, and the investigator is then not clear about the use the dataset that has been collected, and change their mind about the analysis and get frustrated as the data that has been collected can't really answer the question they originally had. Therefore, it comes down to coming up with an agreed plan early on and sticking to the statistical analysis plan. Unless it is a large complex clinical trial, most of the analysis can be straightforward and done in a matter of days. I'm very keen on developing statistical analysis plans very early in the study and sometimes request that the statistician commit in writing the analysis plan.

5. **Do you find it difficult to find a statistician who has knowledge and expertise in the particular field that you are working on?**

Biostatisticians are in relatively short supply, and finding one with therapeutic area expertise is even more challenging. I may be lucky, but I have not struggled to find a statistician who can get the job done. Sometimes, the statistician may be busy with other trials and studies, and may not be able to commit the time and resource, but I am often able to find someone who can take on the task. I find that if you have a good working relationship with one statistician, that often means that you end up working with that person on other studies, as you learn how to work effectively with the person, and there is an element of trust built up. The statisticians also do that: Although they may start with a variable spectrum of clinician collaborators, they eventually identify and work with a smaller group, towards more productive and quality collaborations. In most cases, it is OK to perpetuate those relationships.

6. **Are there any specific knowledge or skill sets that biostatisticians should acquire to raise the standard of collaborative input?**

Having experience in therapeutic areas is an added advantage. I have worked with statisticians who are able to articulate themselves well, and this has not really been a limitation. There is a spectrum of personalities among statisticians I've worked with, although the bias is more towards introverts. For big trials that go on for a long time, there are many people involved in the collaboration from multiple sites, the statistician can play a senior leadership role in the team and that can add tremendous value to the team and collaboration. A statistician should have good people management skills in addition to proactively adding to discussions and leading staff in the collaboration, specifically the data management team, and email site investigators in the team regarding submission of data. It would be really helpful for statisticians to have the diplomatic skills to communicate with other investigators without offending them. It is also helpful for statisticians to gain skills in presentation, so that they can present at investigator's meetings, conferences and other important events.

11.8.3 Interview with Professor John Augustus Rush, MD, AB

Professor emeritus
National University of Singapore
Singapore
12 May 2014

1. **In the past 5 years, how many biostatisticians or data analysts have you worked with?**

   Five.

2. **In what ways have biostatisticians added value to your research projects?**

   They are critical in helping to design the study to ensure the primary aim and secondary aims are met in the most efficient manner. They provide an objective perspective that often helps me separate the forest from the trees and forces me to be decisively clear about what the study is all about. Their independent logically based perspective is invaluable.

3. **Can you describe some of the challenges you may have faced in collaborating with biostatisticians?**

   The main problems can be an overly rigid view of how things should be done without reading what the investigator is trying to accomplish. Some biostatisticians suffer from cognitive inflexibility. The second problem is hearing what the problem is before answering. This can be very troubling especially if the person is not asking clarifying questions. Then he/she will not provide the much needed logical perspective as he/she will not have understood the issue to begin with.

4. **Are there any specific knowledge or skill sets that biostatisticians should acquire to raise the standard of collaborative input?**

Many are inherently shy people. They are often very smart. Sometimes, they do not realise what they say affects the relationship they are needing to develop to collaborate. Useful to have a senior person watch them as they begin to work with collaborators to advise on what the transactions are like. Also need to be curious about in order to understand the needs of the collaborator. Need some EQ enhancement experiences. Sometimes, collaborators are difficult themselves. How to handle grumpy or narcissistic collaborators would be good to know. This knowledge is often from more experienced stats persons. Get a mentor who has been there.

5. **What are the resources available to statistician to gain these skills you've mentioned?**

Some books on conflict resolution may be helpful like 'Getting to Ye's' or even 'How to Make Friends and Influence People'. There may even be some web-based courses.

11.8.4 Interview with Professor John McNeil, AM, MBBS (Adelaide), MSc (London), PhD (Melbourne), FRACP, FAFPHM

Head of School of Public Health and Preventive Medicine
Monash University
Australia
5 August 2015

1. **In the past 5 years, how many biostatisticians or data analysts have you worked with?**

   There has been a lot, and they varied enormously from graduate students who have done a course in statistics to those who have experienced skills and background such as yourself. It's also important that we see data analysts as different from biostatisticians. Data analysts are able to access a database and extract data and generate reports for us, whereas we look to biostatisticians to analyse data in an intelligent and safe way. All of us know the power of modern computers and statistical software, and it is too easy to push the button and not know if it was the right method to use.

2. **In what ways have biostatisticians added value to your research projects?**

   The statistics I use is the 'BOT' (i.e. the 'bloody obvious test'). I can look at it and know what it means. However, increasingly in publications in medical research now, it's not just the use of basic tests, but we have to do more complex things like adjust for variables, etc. Take the ASPREE (Aspirin in Reducing Events in the Elderly) study, for example, where it's not just simply comparing the aspirin versus the 'no aspirin' groups, but figuring out what one needs to do when there is an imbalance after randomisation. Biostatisticians are also going to be much more useful with registry data, where there is a need to risk-adjust, and the biostatistician needs to work with the clinician and epidemiologist to decide, in particular, which variables should be adjusted for (e.g. variables that may be related to the outcome that the clinician can't control), and sometimes the clinician only has partial control (e.g. 50%, 70%) over these variables. I rely on the biostatistician to provide high-level statistical support for the registries, including things like risk-adjustment, control charts and forest plots. What I look for from biostatisticians is not just analyzing data, but finding innovative ways to present and display data.

3. **Can you describe some of the challenges you may have faced in collaborating with biostatisticians?**

   Sometimes clinicians don't understand the statistical approach used in an analysis, particularly for observational studies. For example, in a multi-variable regression model, sometimes several models are presented, each adjusting for a number of covariates. Technical details aside, I often would like to find out the rationale for adjusting for those variables. Sometimes, it may not make sense as they are part of the causal pathway, and particularly in registry data, I would like to find out what variables were risk adjusted for, etc., as these sorts of decisions should be made jointly with the clinicians.

4. **Are there any specific knowledge or skill sets that biostatisticians should acquire to raise the standard of collaborative input?**

   My wish list is for biostatisticians to have a broader understanding of demography and modelling, so as to allow for the skills to overlap. For instance, we are no longer interested to know from a study that the hazard ratio is 0.8 and that it is statistically significant, but we also want to know how many people are going to be living and for how much longer. We want to turn the results into something useful for the public and the politicians. I think biostatisticians often stop at a level which is increasingly half-way along the analysis pathway, and they should scale up. I would also like biostatisticians to find new and innovative ways to present data (e.g. be able to use all the latest graphics packages) and be able to advise on a choice of possible methods.

5. **What are the resources available to statisticians to gain these skills you've mentioned?**

   If anyone comes to me and says that he/she would like to go for a course, work in a unit for a couple of weeks or go to a meeting, it is highly unusual for us to say no, unless the budget is in dire straits. Biostatisticians, working on registry data for example, should be given opportunities to present in major meetings on topics such as 'Challenges in analyzing registry data'.

11.9 Conclusion

1. The main problem faced in collaboration relates to the long time taken to write up a manuscript or grant proposal.

2. The primary source of training on collaboration/consultation skills is from other colleagues.

3. Satisfaction with current skills is related to obtaining training from collaborators.

4. Data analysis ranks the highest in terms of frequency of tasks performed, but surprisingly data management is the second most frequent.

5. Improved communication between biostatisticians and clinicians as well as better matching of biostatistics skills with projects are ways to enhance greater collaboration with clinicians.

6. Clinicians generally report a healthy relationship with statisticians, but better communication is also suggested as a way to improve collaborations.

Appendix: Sample Survey Questionnaire

1. **What is your highest level of educational qualification?**
   a. PhD
   b. Master's degree
   c. Bachelor's degree

2. **Who is your primary employer?**
   a. Hospital/health care institution
   b. Academic institution
   c. Private sector institution

3. **How many years have you been working as a biostatistician/data analyst?**

4. **In the past 12 months, how many collaborative projects have you been/are currently involved in as a biostatistician/analyst?**

5. **In the course of collaborations with clinicians/other health researchers, what are the problems you have ever faced? Check all that apply.**
   a. Difficulty in understanding medical terms
   b. Difficulty in explaining statistical concepts
   c. Unrealistic deadlines set by collaborator
   d. Collaborator approaches me at the last minute
   e. Collaborator engages in data dredging
   f. Authorship issues
   g. Collaborator takes a long time to write up manuscript/grant proposal
   h. Nil
   i. Others (please specify)

6. **How satisfied are you with your current skills in collaborating and in consultations with clinicians/health researchers?**
   a. Completely dissatisfied
   b. Dissatisfied
   c. Somewhat satisfied
   d. Very satisfied
   e. Completely satisfied

7. **Where have you obtained training on collaboration/consultation skills? Check all that apply.**
   a. Formal school course
   b. Books/journals

   c. Learning from other colleagues

   d. Structured on-job training

   e. Websites and other electronic resources

   f. From collaborators themselves

   g. Nil

   h. Others (please specify)

8. **In a typical month, how often do you perform the following tasks for collaborative projects? (Never/almost never, seldom, often, always.)**

   a. Data analysis

   b. Statistical analysis plan

   c. Sample size/power calculation

   d. Hypothesis formulation

   e. Data management

   f. Randomisation

   g. Questionnaire/form design

   h. Others (please specify)

9. **In your opinion, which of the following should be addressed to enhance greater collaboration with clinicians? Check all that apply.**

   a. Communication between biostatisticians and clinicians

   b. Accessibility and availability of biostatisticians

   c. Technical knowledge of biostatisticians

   d. Increasing future pool of trained biostatisticians

   e. Providing better matching with specific biostatistics skill that may be required for project

   f. Providing institutional requirement that all clinical research projects should include a biostatistician collaborator

10. **Please rank the following skills/knowledge most important for you to gain in order to improve on collaborations with clinicians (1 – Most important and 5 – Least important)**

   a. Advanced statistical models

   b. Time management

   c. Project management

   d. Communication skills

   e. Understanding medical terminology

Key Learning Points

1. Longer time taken to write the manuscript and approaching the biostatistician at the 11th hour are the main problems identified when collaborating with clinicians.

2. The top course of learning collaboration/consultation skills is from other colleagues.

3. The most frequently performed task for a collaborating biostatistician is data analysis, but data management comes in a surprising second.

4. Communication has been identified as the key issue to address in order to enhance greater collaborations by both biostatisticians and clinicians.

5. Clinicians are unanimous in their views that biostatisticians clearly add value to research projects in various ways.

References

Centre for Quantitative Medicine. http://www.duke-nus.edu.sg/research/clinical-sciences/centre-quantitative-medicine-cqm#member-profile. Date accessed: 5 March 2014.

Surveymonkey. https://www.surveymonkey.com. Date accessed: 5 March 2014.

Index

Note: Page numbers followed by f and t refer to figures and tables, respectively.

Printed in the United States
by Baker & Taylor Publisher Services